1/16/14
$39.95

Hunger, Thirst, Sex, and Sleep

Hunger, Thirst, Sex, and Sleep

How the Brain Controls Our Passions

John K. Young

ROWMAN & LITTLEFIELD PUBLISHERS, INC.
Lanham • Boulder • New York • Toronto • Plymouth, UK

Published by Rowman & Littlefield Publishers, Inc.
A wholly owned subsidiary of The Rowman & Littlefield Publishing Group, Inc.
4501 Forbes Boulevard, Suite 200, Lanham, Maryland 20706
www.rowman.com

10 Thornbury Road, Plymouth PL6 7PP, United Kingdom

British Library Cataloguing in Publication Information Available

Library of Congress Cataloging-in-Publication Data Available

ISBN 978-1-4422-1823-9 (cloth : alk. paper)
ISBN 978-1-4422-1825-3 (ebook)

∞™ The paper used in this publication meets the minimum requirements of
American National Standard for Information Sciences—Permanence of Paper
for Printed Library Materials, ANSI/NISO Z39.48-1992.

Printed in the United States of America

Contents

Acknowledgments

$\mathscr{F}$irst and foremost, I want to thank my wife Paula for many years of loving companionship and encouragement. I also want to thank my brother Bill for his encouragement and support. Second, I want to acknowledge many stimulating and enjoyable conversations about the science of the brain that I've had with colleagues such as Thomas Heinbockel, Jim McKenzie, and Blair Turner here at Howard University.

OTHER BOOKS BY JOHN YOUNG

Sacred Sites of the Knights Templar. (2003). Fair Winds Press: Gloucester, MA.
The Building Blocks of Human Life: Understanding Mature Cells and Stem Cells. (2007). Recorded Books: NY.
Human Anatomy: The Beauty of Form and Function. (2008). Recorded Books: NY.
Introduction to Cell Biology. (2010). World Scientific Press: Singapore.

Introduction

$\mathscr{I}$ feel compelled to begin this book by offering you readers a brief apology. I have been studying and thinking about the hypothalamus for more than 30 years, and after that long a period, I probably am a bit "over the top" about this subject. As the saying goes, if you are a hammer, everything looks like a nail! Since I am obsessive about the hypothalamus, it is probably not too surprising that when I look for the sources of human happiness and human problems, for me, parts of the hypothalamus are what come to mind first.

It is not too bad a sin to focus so consistently on a part of the brain that is so interesting. As I hope I will show you, the hypothalamus does have a profound influence on many important portions of our lives: it commands us when to eat or drink, it tells us when we need to sleep, it regulates our sex lives, and it controls many of the hormones that control our bodies. Abnormalities in the hypothalamus appear to play a role in many medical conditions, including disorders that you would not expect to be related to the brain, like diabetes and cancer. Also, for neuroscientists, the hypothalamus is an enticing area of the brain because it is neither too complicated to completely understand nor too simple to be boring.

Some parts of the nervous system perform elegant analyses of data that we still can scarcely comprehend. The cortex of the brain, for example, processes information in a complex way that lets us identify faces as human, distinguish the faces of our friends from those of strangers, and guess the emotional state of our friends from their expressions. Recent research has begun to show how nerve cells in one brain area (the hippocampus, located just beneath the temporal cortex) allow us to recognize our friends and acquaintances. For example, researchers in California recently studied electrical

1

activity in single hippocampal nerve cells in patients that had been implanted with brain electrodes to control epileptic seizures. They found single nerve cells in this region that reacted to a picture of Marilyn Monroe! These same nerve cells reacted to Monroe's voice or the mention of her name but did not react to the features of other people.[2] In another study of the temporal cortex of monkeys, researchers identified clusters of nerve cells that preferentially react to faces but don't respond when a monkey is shown a picture of a hand or a leg.[1] Neuroscience research like this suggests that vast amounts of information on the shape of faces, the sounds of voices, and the identity of persons must be being analyzed by millions of neurons that converge onto only a few cells that enable the brain to make a neural representation of the characteristics of only one individual. How this complicated process actually takes place is still something of a mystery.

Other parts of the nervous system supervise simpler tasks. Circuits within the spinal cord stimulate the flexor muscle of the arm (biceps brachii) to contract when we lift a box, and at the same time relax the opposing extensor muscles—the triceps—so there is not undue resistance to arm movement. This relatively simple coordination of extensor and flexor muscles is easier to understand and model than the tasks overseen by the cortex. Likewise, circuits a bit higher up in the brain (the medulla, just above the spinal cord) provide rhythmic stimulation to the diaphragm to maintain breathing or regulate the heart rate. These regulatory influences upon muscle groups are gradually yielding their secrets to contemporary neuroscience.

The hypothalamus falls between the elegance of the cortex and the simple mechanisms of the spinal cord. The hypothalamus does not generate thoughts or analyze words or numbers, but neither is it just a simple mechanism that keeps muscle groups in balance with each other. The hypothalamus oversees the overall health of the body by maintaining proper blood levels of nutrients, hormones, salts, and fluids. It contributes to the basic drives that underlie sexual behavior, feeding, drinking, and arousal. It controls body temperature and regulates the responses of the autonomic nervous system to stresses. Many of the basic pleasures and pains of life are subject to the influence of this part of the brain. In the last decade or so, the cells and pathways that regulate these features of our lives have finally become better understood. In this book, I aim to show you how some of these hypothalamic functions are accomplished and how they relate to a number of fascinating medical conditions.

Some of the claims and statements that I have made in this book may be confusing or simply hard for you to believe. In an attempt to convince you that my fascination with the hypothalamus is based on reality, I have tried to provide numbered references to the scientific literature after each chapter.

These scientific articles may be easily accessed on a website provided by the National Library of Medicine called www.pubmed.gov. Simply type in the name of the author and the date, and an abstract of the article (or often the article in its entirety) pops up. You might like to start by looking at the papers listed for this introduction in the Notes section of this book (p. 145).

The Anatomy of the Hypothalamus and the Control of Hunger

*W*hen I first started to think about becoming a scientist, I knew almost nothing about the hypothalamus. I was only 21 years old and didn't really know the direction I would take in science. I had just been admitted to the graduate program in the Department of Anatomy at UCLA, but I was nervous about meeting the professors and wasn't sure what kind of research I would want to do. All I knew was that I had been drawn to study the comparative anatomy of animals ever since I was a teenager.

My interest in anatomy came about because I was nuts about dinosaurs. That wasn't terribly special—lots of children spend their spare time thinking about *T. rex* or *Triceratops*—but I carried mine to extremes by checking out all the books on dinosaurs that I could and by memorizing all of their complicated Latin names and body parts. So, by the time I was older, I already had some understanding of how the bones and muscles of dinosaurs were put together. This is one way to get interested in anatomy as a profession.

I initially wanted to continue my dinosaur mania by trying to become a paleontologist, but I soon learned that this possibility was pretty remote. In fact, six years previously I had briefly visited a paleontologist in his office at the LA County Museum of Natural History, where all the specimens of saber-toothed tigers and mastodons from the famous La Brea tar pits were stored. I wanted to ask him what a career in paleontology would be like. He did his best to startle me by going over to a wall and extracting a huge wooden drawer from a stack of drawers. Inside were hundreds of identical looking cylindrical bones.

"Do you know what these are?" he asked me. I could only shake my head. "These are the penis bones from all the dire wolves we collected from the tar pits. Quite a collection, don't you think?" Shy and flustered, I couldn't think of a good way to respond to this question! When I finally got around to asking him how I, too, could become a paleontologist, he said his best advice would be to study hard and then wait for someone in the field to die. There just weren't enough jobs in paleontology for me to reasonably expect to be hired in the next 20 years! This was rather discouraging, but at least it helped me set more reasonable goals. If I couldn't become a paleontologist, perhaps I could become an anatomist. So that's how I happened to appear at UCLA in 1972.

I was welcomed into the department by Dr. Charles Sawyer ("Tom" Sawyer to his friends), who was an eminent endocrinologist. Dr. Sawyer had me test the waters right away: he put on his lab coat and walked me down the hall from his office right into the gross anatomy lab to show me the cadavers I would be dissecting in my first classes. Numerous embalmed bodies, wrapped in gauze to keep them moist and stored in stainless steel dissection tanks, filled the lab. This was my first experience viewing dead bodies; I was surprised that the embalming process had turned the skin color of all of them to the color of paper grocery bags. Fortunately, I didn't disgrace myself in front of Dr. Sawyer by getting upset at the panorama of death around me and felt pretty confident that I could master the upcoming course work. Over the next few months, I would learn to cut into the dead tissues and identify all the nerves, muscles, and vessels. However, what kind of research would I want to do? What would a career as an anatomist mean to me?

Like most young people, I wanted to do something important and perhaps study something that might lead to a cure for some medical condition. One condition—the feeding disorder called anorexia nervosa—was attracting considerable attention at the time. Anorexia is a fascinating condition in which people voluntarily starve themselves to achieve incredibly low body weights. It affects many people of all walks of life, including, for example, the singer Karen Carpenter, who succumbed to anorexia at about the time I became interested in the disorder. It seemed amazing to me that anyone could lose so much weight and keep it off by dieting, when almost everybody else (me included) could hardly manage to lose five pounds and keep it off for more than a year. What controls our appetite for food, anyway? I was lucky to have been admitted into a department where a study of the hypothalamus, the part of the brain mainly responsible for the control of hunger, was a focal point of research. It was not long before I too began to learn about the anatomy and function of the hypothalamus so that I could better understand what might cause anorexia nervosa.

WHERE IS THE HYPOTHALAMUS, AND WHAT DOES IT LOOK LIKE?

To learn how the hypothalamus works, the first thing I had to do was to understand where the hypothalamus is, what it looks like, and how it is organized. The hypothalamus is located in a very inaccessible place at the very bottom of the brain, deep within the skull. To gain access to the hypothalamus and show it to the students that I have trained over the years, I have frequently had to first remove the brain from the skull. This is not an easy task.

Fresh brain tissue is very delicate (not unlike congealed oatmeal in consistency). Getting it out of the skull requires several careful steps. First, cuts are made through the top of the skull and the base of the skull with an instrument called a Striker saw that allow the skullcap to be removed. Next, the membranes covering the brain (meninges) are incised, and then the brain is carefully removed by severing its connection to the spinal cord. Finally, the brain is immersed in a solution called formalin.

Formalin is a very useful chemical for anatomists. It is created by forcing bubbles of a gaseous compound called formaldehyde into water. This is not unlike forcing carbon dioxide into water to create soft drinks: the carbon dioxide dissolves in water, particularly in solutions that are kept under pressure. When the pressure is decreased by opening a can of soda, the carbon dioxide is less soluble and escapes from the solution as bubbles. Formaldehyde is more soluble than that, and can reach concentrations of 37% in freshwater. These solutions are much too acid, however, to be used to preserve tissues. Instead, formaldehyde solutions are diluted 1:10 in water buffered with sodium phosphate to make formalin. This chemical coagulates proteins, makes brains firmer and more rubbery, and prevents bacteria or enzymes from degrading the dead tissue. It also is rather foul smelling and very irritating to the nose, so that fixation of brain tissue with formaldehyde is best performed within a fume hood that sucks all the toxic vapors up a ventilation shaft.

After a couple of days in formalin, the brain is firm enough to be examined for the landmarks that identify the hypothalamus.

X MARKS THE SPOT

Many of the anatomical terms we now use to describe the brain date from a surprisingly long time ago. Some of these originated from the writings of a Dutch anatomist, Andreas Vesalius (1514–1564), who published some of the first diagrams of the brain. Vesalius distinguished himself from other scientists of his era by actually going to the trouble of performing dissections on actual human cadavers instead of relying on information passed on from books written in previous decades or even in previous centuries. By doing direct scientific work himself, Vesalius was able to correct many long-standing misconceptions about anatomy (e.g., the human heart contains two ventricles, not three!).

In one of the diagrams in Vesalius' famous book, a large X-shaped structure is visible on the underside (ventral surface) of the brain. This is the optic chiasm, the place where the two optic nerves from the eyeballs move toward the midline and fuse together before diverging once more to progress backward toward the visual cortex. This place where the optic nerves cross forms the anterior border of the hypothalamus (Fig. 1-1).

The posterior boundary of the hypothalamus is marked by two small spherical swellings, which some naughty anatomist named the mammillary bodies, apparently after some other anatomical preoccupation of his. The chunk of brain tissue located between these boundaries is not much bigger in diameter and depth than a 50-cent piece, but nevertheless it has an influence upon brain function that is all out of proportion to its size.

HUNGER AND THE HYPOTHALAMUS

The sensation of hunger for food is one of the strongest impulses we feel in day-to-day life. Normally, hunger is only a transient and nontraumatic experience, because we tend to satisfy it quickly before it can dominate our lives. However, when people are not so fortunate as to have a ready supply of food, the intense, driving qualities of real hunger are unmasked.

One well-studied example of how hunger can dominate the lives of thousands of people took place during World War II in the western regions of the Netherlands. During the so-called Hunger Winter of 1944–1945, food supplies became suddenly scarcer, and food was rationed to supply only 600 calories per day instead of the normal 2000 calories. As a result, almost 18,000 people starved to death, and the remainder of the population experienced severe hunger. The psychological effects of hunger were carefully documented during this period. Starving people lost interest in almost all topics that were not related to food, and they thought and dreamed about food continuously. They experienced a loss in sexual appetite, became depressed, and in some cases experienced hallucinations or a worsening of previous psychological maladies. Not surprisingly, most of these symptoms promptly disappeared when food became available again.[27]

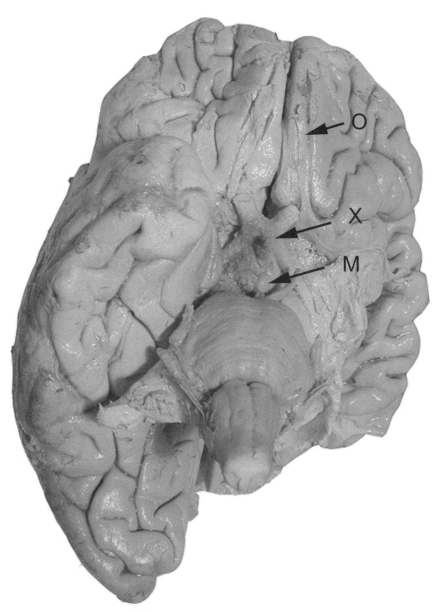

Figure 1.1. A photograph of the underside (ventral surface) of the brain, showing the optic chiasm (X) and the mammillary bodies (M) that mark the anterior and posterior boundaries of the hypothalamus. One important anterior structure that sends information to the hypothalamus is the olfactory nerve and bulb (O in this photograph). Just posterior to the mammillary bodies is the ventral surface of the pons, which is part of the midbrain. The pons sends information to the medulla and spinal cord at the very bottom of this picture. *Photo courtesy of Dr. Blair Turner, Dept. Anatomy, Howard University*

What is the origin of the sensation of hunger? We all experience hunger as a growling of the stomach and the secretion of saliva into the mouth. These sensations, however, are most likely symptoms of hunger but not causes of hunger.

Hunger is obviously a response of the body to a lack of calories. But how does the body "know" that we have run out of fuel? What organs monitor our overall state of nutrition? At the turn of the century, many scientists thought that the contractions of an empty stomach (we know these as "hunger pangs") could be an important signal that we need food. However, by the 1940s, this line of thinking was discredited when it was discovered that injections of insulin, which make a dog very hungry, do not cause stomach contractions at all until the animal has actually eaten some food. So, an empty stomach does not provide an essential signal that we need to eat. A full stomach, of course, does depress our appetite; if a water-filled balloon is chronically placed in the stomach of a dog, the dog will eat less and gradually lose weight.

However, much of the input to the brain from gastrointestinal organs can be eliminated without a loss in the ability to maintain a normal body weight. For example, in the 1950s, a common approach for the treatment of stomach ulcers was to cut the vagus nerve that connects the brain to the stomach. This not only decreased the secretion of stomach acids and helped to cure ulcers, but it also prevented any sensory information from the stomach from reaching the brain. Nevertheless, such ulcer patients were still able to control their food intake and maintain a normal body weight, so these types of studies pointed to the brain as a more fundamental regulator of feeding.

The control of feeding and obesity is not only interesting to me and you but has acquired much more medical importance than in the past. In some counties of the United States, almost 30% of the population can be judged to be obese, which shows that we are all getting fatter than we were in previous decades.[8] Why is this happening? And more to the point, how can it be avoided? Obesity greatly increases the risk for damaging diseases like diabetes and hypertension. So, a better understanding of why we are eating more and becoming fatter would be important for many reasons.

HYPOTHALAMIC DAMAGE CAUSES HUMANS TO GET VERY FAT

The first indication that the hypothalamus could be involved in the control of hunger came from a report published in 1901 by an Austrian physician named Alfred Fröhlich. This paper described a patient who had acquired a tumor of the pituitary that had encroached upon the hypothalamus and damaged it. This patient became much fatter than normal and also suffered from infertility. Subsequently, these symptoms were found in additional patients, and this combination of symptoms was termed Fröhlich's syndrome in Alfred Fröhlich's honor.[19] These data showed that damage to the hypothalamus does seem to cause abnormalities in appetite control. However, which parts of the hypothalamus were involved, and how did they normally sense that the body needed more calories?

Detailed information about the hypothalamic nerve cells that control hunger was just not available in Fröhlich's time. Techniques for studying the anatomy of the brain at the cellular level had only just been worked out at the end of the 19th century. These techniques required that the brain be cut into thin sections, followed by mounting the sections on slides and staining them to view the nerve cells under the microscope. This allowed scientists to begin to understand how the hypothalamus is organized and to see which parts of it are most important for the control of hunger.

Cutting brain sections has turned out to be a very common activity for me. First, I take a chunk of brain tissue that needs to be studied, place it onto a flat piece of metal, and pour powdered dry ice over it. This freezes it solid so that it can be cut into relatively thin sections. These sections are only about 40 microns (40/1000 of a millimeter) thick, about the thickness of a piece of tissue paper.

To cut a brain into sections that are this thin, the flat metal plate bearing the brain tissue is slid under the edge of a very sharp steel knife that is locked into an apparatus called a microtome. Each time I shove the metal plate forward, a gear mechanism moves it 40 microns higher up. Thus, by moving the plate back and forth beneath the knife, I can cut a series of sections. As the brain is cut, not unlike cutting a salami into sections for a sandwich, all the sections are deposited into a buffer solution. There, each section unfolds again in the liquid. Finally, each section can be retrieved from the buffer with an artist's paintbrush, placed onto a drop of water on a glass slide, and laboriously spread out to dry down onto the slide. I usually play music in the background while I'm doing this to soothe my nerves during this tedious part of my job. Subsequently, each section can be stained with a number of stains (a blue stain called thionine is a common one) to reveal the collections of nerve cells seen under the microscope.

Techniques like these show that nerve cells are not distributed uniformly throughout the hypothalamus but are organized into clusters of neurons termed hypothalamic nuclei. This rather confusing term does not refer to the nucleus of an individual cell but to a concentration of thousands of cells into a structure that often is just large enough to be seen with the naked eye. It's important to understand the function of each of these clusters of nerve cells, since each nucleus in the hypothalamus is connected via specific pathways to other parts of the brain, and since each cluster contributes to hypothalamic function in its own special way.

When a brain is cut in a cross section at the level of the optic chiasm, the basic components of the brain can be seen (Fig. 1-2). The top of the brain is formed by the cerebral cortex, where complex computations about movement, language, and sensations are performed. The two hemispheres of the cerebral cortex communicate with each other via large bundles of nerve fibers called commissures. The top commissure (corpus callosum) communicates information between the left and right parietal and frontal lobes of the cortex. If this

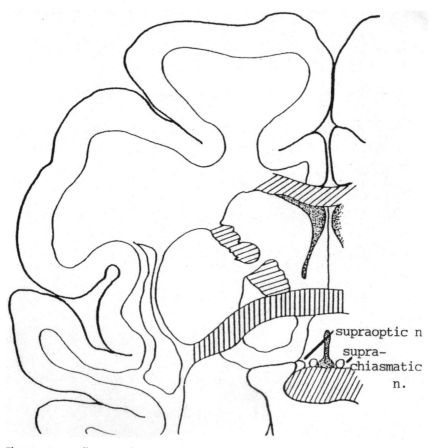

supraoptic n

supra-
chiasmatic
n.

Figure 1.2. A diagram of a cross section through the part of the brain that contains the anterior portion of the hypothalamus. Bundles of nerve fibers (commissures) that communicate between the two hemispheres of the brain are shaded in diagonal lines. A fluid-filled cavity called the third ventricle is located between the right and left halves of the hypothalamus.

communicating avenue between each side of the brain is damaged or cut, the two halves of the brain lose their connections and patients will suffer peculiar symptoms. In these "split brain" situations, information from the right eye does not reach the left brain, which controls speech. Patients with this condition, when shown an object using the right eye, can often not speak about it, because the speech centers of the left cortex never learn about it, but they can pick the object up and manipulate it completely normally with their right hand.

Another commissure, which forms the roof of the anterior hypothalamus, is called the anterior commissure and sends information between the two temporal lobes of the cortex. A final bundle of nerves, on the bottom of the brain, is the optic chiasm, which forms the floor of the anterior hypothalamus.

The hypothalamus itself is divided into two equal halves by a fluid-filled cavity called the third ventricle. On each side of the anterior hypothalamus, two major nuclei can be seen: the supraoptic nucleus and the suprachiasmatic nucleus. These collections of nerve cells, as we shall see, are critical for the control of water content in the body and for the generation of 24-hour-long cycles in many brain activities.

Another cut through the brain taken about a centimeter further back would slice through additional hypothalamic nuclei. At this more posterior location, the optic chiasm has divided into two optic tracts, and the third ventricle is larger. On either side of the third ventricle lies a nucleus which was termed the paraventricular nucleus (from the Latin word *para*, which means "alongside," since it is located alongside the ventricle). Below that, another nucleus called the ventromedial nucleus is visible. Even lower, a smaller nucleus called the arcuate nucleus is visible. It has only been in the last 10 years that the critical influence of this relatively small cluster of cells upon hunger and obesity has been understood. Finally, the bottom of the hypothalamus at this brain level is formed by a structure called the median eminence. This forms an attachment point for the pituitary gland, which hangs suspended from the bottom of the hypothalamus.

The posterior portion of the hypothalamus is mainly occupied by large, circular nuclei called the medial mammillary nuclei, which produce the spherical swellings known as the mammillary bodies.

A cross section through the huge human brain is too large to be portrayed very well in a single photograph. However, the brains and hypothalami of mice are smaller and still show very much the same features as does the human hypothalamus (Fig. 1-3). Because of the anatomical similarities between the hypothalami of rodents and humans, studies of the functions of the rodent hypothalamus can shed a lot of light on the function of the human hypothalamus.

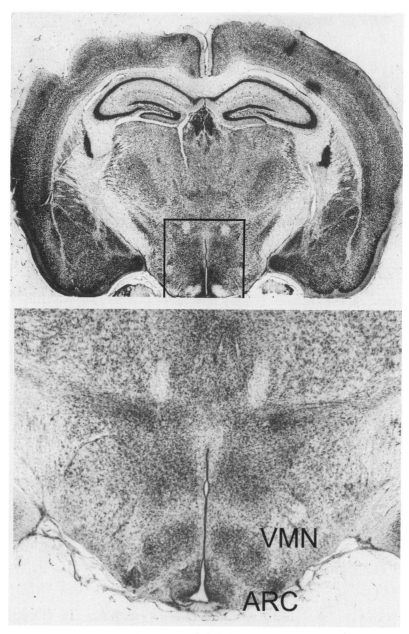

Figure 1.3. Top: a cross section of a mouse brain, showing the cortex, thalamus, and (enclosed by a square) the hypothalamus. Each tiny dot in this picture represents a single neuron, showing that there are hundreds of thousands of neurons visible in each and every section of the brain. Two highly stained layers of neurons just beneath the cortex constitute the hippocampus, a part of the brain important for learning and memory. Bottom: higher magnification view of the hypothalamus, showing the ventromedial and arcuate nuclei and the median eminence.

The nuclei named previously are not the only nuclei present in the hypothalamus. Some of these additional nuclei—for example, the sexually dimorphic nucleus and the tuberomammillary nucleus—will be discussed in later portions of this book. Other nuclei, such as the dorsomedial nucleus (just above the ventromedial nucleus) have been far less studied than the others and will receive no attention in this book. One nucleus, the lateral tuberal nucleus, is unique in that it is present in the human but absent from hypothalami of other animals and thus has received virtually no study as to its function.[38]

The only sensible way to find out how the hypothalamus controls appetite is to study feeding behavior in experimental animals and see if damage to the hypothalamus can provoke symptoms similar to those seen in Fröhlich's syndrome. The animal of choice for studying feeding behavior is the lab rat.

STUDYING THE HYPOTHALAMIC
CONTROL OF APPETITE IN RATS

Rats, in my opinion, have acquired a bad reputation that they do not deserve. To be sure, wild rats can be pretty aggressive and can spread disease. I have found that albino lab rats, however, are affable little creatures who can be docile, clean, and even affectionate. Moreover, it is pretty easy to study their feeding behavior. In my lab, each rat is given a daily ration of rat chow, which is supplied by Purina company as a fine powder containing a balance of nutrients and vitamins that is just right for a rat. Some researchers use specially constructed metal feeding dishes to contain the chow, but I prefer used baby food jars, which are just the right size and which have smooth glass surfaces that can't cut the nose of a rat trying to get a meal.

Rats usually sleep throughout much of the day, but during the night they will walk over to the feeding jars in their cages and eat small meals throughout the evening. I've spent many days and nights filling baby food jars, putting them in cages, and weighing them after specific intervals to see how much has been eaten. Even if the rat spills a little food each night, the spillage can be collected on paper towels and weighed to correct any errors.

Some of my studies have examined the effects of hormones on feeding behavior. For example, an injection of a small amount of insulin under the skin of a rat causes blood sugar concentrations to fall slightly. The hypothalamus reacts to a fall in blood sugar by causing the rat to eat more chow over the subsequent several hours. Other hormones, like estrogen, have a depressive effect upon appetite. These effects can be easily monitored by simply weighing food dishes at the right times.

To see how the hypothalamus regulates feeding, early researchers created small areas of damage in the hypothalamus and measured feeding behavior. A standardized procedure for doing this was worked out during the 1940s by Hetherington and Ranson at Northwestern University in Chicago.[19] After giving a rat a heavy dose of anesthesia, the head of the unconscious rat would be positioned firmly in a stainless steel frame called a stereotaxic apparatus (larger ones were also developed for human brain surgery). Then a small hole would be drilled in the skull with a dental drill similar to the ones used on people to excavate a tooth cavity. Finally, a very thin platinum wire would be lowered into the brain. The wire would be completely covered with an insulating layer of epoxy except for the final 0.3 mm at the tip of the wire. This would permit a tiny spark to be generated at the very tip of the wire after 1 milliampere of electricity was passed through it. This is a very small amount of electricity; house currents used to operate lightbulbs and TVs usually employ 15 amperes of electricity, over 1000 times as much as is used in research on brain lesions.

The minute tissue damage generated by such electrolytic lesioning procedures could be less than a cubic millimeter in size, but after the rat awoke from the anesthesia and recovered from the surgery, it often would display dramatic changes in behavior. Rats with lesions near the third ventricle and the bottom of the hypothalamus (ventromedial lesions) would never seem to feel full and would eat continuously if offered tasty food. They would gradually become so obese that they had difficulty walking around their cages. On the other hand, lesions located far away from the midline (lateral hypothalamic lesions) produced dramatic decreases in eating and drinking, so that rats became very thin. These types of data led to proposals that the ventromedial hypothalamus constituted a type of "satiety center" and that the lateral hypothalamus formed a so-called "feeding center."

These experiments did focus attention upon the hypothalamus and the control of eating, but they quickly attracted a number of criticisms. For one thing, electrolytic lesions do not only destroy nerve cells in the hypothalamus, but they also interrupt bundles of axons from other parts of the brain that form pathways through the hypothalamus. For example, a major pathway that travels through the lateral hypothalamus, called the median forebrain bundle, sends information from anterior (olfactory) brain regions to centers in the brain stem. Other pathways could also be interrupted by hypothalamic lesions. Perhaps the changes in behavior caused by lesions could partly be explained by disconnections between different parts of the brain rather than by damage to hypothalamic nerve cells. Maybe the hypothalamus was just a junction for all of these circuits and didn't control feeding by itself at all.

How could these criticisms be answered? And even if hypothalamic nerve cells did turn out to control hunger, which specific cells were important, and how did they function? If a "satiety center" does exist in the hypothalamus, how does it acquire information about how many calories are present in the body? One could imagine that pressure sensors on the bottom of the feet could keep the hypothalamus informed about how fat a person is, but this crude measurement of obesity is not a likely mechanism for feeding control. How does the hypothalamus "know" that a person is becoming fatter so that appetite can be restrained?

More accurate answers to all of these questions required advances in neuroscience techniques. Since the 1940s, scientists have learned to identify specific subpopulations of neurons within a given brain region and have found where these neurons send information. Also, scientists have begun to understand the signals that inform the hypothalamus about the level of fat in the body.

BLOOD-BORNE SIGNALS OF THE BODY'S CALORIES

There are a number of obvious candidates for signals from the body to the brain that communicate whether or not we are lean or obese. For example, after eating a meal, blood levels of nutrients such as glucose or fatty acids increase; also, meal-related increases in blood hormones such as insulin (from the pancreas) and cholescystokinin (from intestinal cells) could serve as signals for calories. However, how would these be detected by the hypothalamus?

This is not as simple a question as one might think: in most of the brain, capillaries form tightly sealed tubes that do not permit the free diffusion of molecules into nervous tissue. This unusual property of brain capillaries is referred to as the blood-brain barrier. The blood-brain barrier protects the brain from most harmful substances that may be circulating within the bloodstream. To breach this barrier, nutrients such as glucose must be transported right through the cytoplasm of capillary cells via specialized transporter proteins. This is very unlike most of the body, in which the junctions between capillary cells are very leaky and allow the free passage of nutrients into tissues like muscle and fat. So, wouldn't the blood-brain barrier block the entry of nutrient signals into the hypothalamus?

It is probably not a coincidence that specialized capillaries of the arcuate nucleus, and also the nearby median eminence, are much more permeable than most brain capillaries.[4,37] This can be shown quite easily: if a mouse brain is stained for blood-borne proteins such as gamma globulin, it's easy to demonstrate that these blood proteins spill over into the arcuate nucleus from permeable capillaries but don't penetrate into the rest of the brain.

Another peculiar thing about the arcuate nucleus is that cells lining the nearby third ventricle, called tanycytes, extend long processes that form a barrier between the arcuate nucleus and the rest of the brain.[21] The leakiness of arcuate capillaries, plus the barrier that surrounds the arcuate nucleus, causes this nucleus, but not the rest of the hypothalamus, to be freely exposed to circulating molecules. This allows the arcuate nucleus to function as sort of an early warning system for the brain, detecting chemicals that freely float around in the bloodstream. Many of these molecules, like glucose, insulin, and fatty acids, decrease appetite when infused into the hypothalamus.[1] This arrangement allows only a restricted part of the hypothalamus the ability to sense circulating chemicals, while preventing circulating chemicals from getting into the rest of the brain and possibly damaging it.

The unique permeability of the arcuate nucleus, which enables it to act as a sensor for blood-borne molecules, also makes it unusually vulnerable to toxic chemicals that can be injected into the blood. Certain kinds of toxic molecules have been used to damage the hypothalamus to see what effects this damage has upon the control of appetite.

A prominent example of such a toxic molecule is goldthioglucose (GTG), a molecule of sugar with a gold atom attached. This synthetic molecule was first studied in the 1950s by researchers interested in treating arthritis. Gold-containing compounds had been found to improve the symptoms of arthritis by blocking attacks upon joints by the immune system. Naturally, drug companies were interested in developing gold-containing medicines that they could potentially sell to treat arthritis. To do this, they had to inject many gold-containing compounds into mice to study the possible toxicity and side effects of these chemicals. When GTG was tested for antiarthritic properties, researchers were astounded to find that some of the treated mice became very fat! What in the world was going on?

It was eventually determined that the GTG was taken up by glucose-sensitive cells in the arcuate nucleus, which responded by dying and leaving behind a pale-staining glial "scar." This type of lesion, like electrolytic lesions, seemed to destroy feeding-restraining neurons and thus altered the hypothalamic control of hunger.[12]

What cells are affected by GTG? It seems likely that GTG does not directly damage nerve cells themselves but is toxic to specialized helper cells called glial cells. What are glial cells?

Approximately 60% of the cells in the brain are neurons, but the remaining 40% of cells are nonneuronal glial cells, which are divisible into a number of different types: (1) astrocytes; (2) oligodendroglia, which make an insulation called myelin that surrounds nerve fibers; and (3) microglia, which clean up debris in the brain shed by damaged or dying cells.[10] Astrocytes themselves account for about 30% of the total volume of most brain areas.[32] For many years, these cells were virtually ignored by neuroscientists and were thought to simply occupy the spaces between neurons. Now, however, it is becoming recognized that astrocytes and other glial cells play essential roles in regulating the functions of neurons and other cells. For one thing, it is very likely that the unusual permeability of capillaries within the arcuate nucleus is due to specialized instructions coming from arcuate astrocytes that dictate that the capillaries become leaky.[37]

Astrocytes of the arcuate nucleus also have unusual glucose transporter proteins in their cell membranes which enable them to take up glucose from the bloodstream in much greater amounts than neurons. The only other places where these specific glucose transporter proteins are found in the body are in other glucose-sensing organs like the liver or the pancreatic islets of Langerhans, which secrete hormones in response to changes in glucose concentrations in the blood.

Normally, these high-capacity glucose transporter proteins allow the glial cells to function as glucose-sensing cells. However, when the cells are tricked to take up a toxic form of glucose (goldthioglucose), these cells die and cause damage to neighboring neurons.[12,36] This is only one example among others (discussed later in this book) of how glial cells contribute to the function of neurons.

Another toxic molecule that damages cells in the arcuate nucleus is an amino acid called glutamate. Glutamate is utilized by many neurons of the brain as a neurotransmitter that activates nearby neurons. It can also be injected under the skin of a rat and will travel to the leaky capillaries of the hypothalamus to bind to hypothalamic neurons. After exposure to glutamate, the nuclei of nerve cells shrink up and darken, and the nerve cells subsequently die. Mice or rats injected with glutamate also thus develop abnormalities in hypothalamic function and become obese. Damage induced by glutamate kills nerve cells in the arcuate nucleus but does *not* interrupt nerve fibers that are merely passing through the area and does not damage nerve cells adjacent to the arcuate nucleus.[35,37] So, these types of studies finally established unequivocally that nerve cells native to the hypothalamus themselves directly influence feeding and obesity. Thus, introduction of toxic molecules into the bloodstream has enabled scientists to perform "brain surgery" on the arcuate nucleus alone without ever making a single incision into the skull of an animal and without damaging other parts of the brain.

One other curious feature of glutamate is that it is much more effective in damaging the hypothalamus in newborn rats than in adult rats, since newborn rats don't metabolize and break down systemically administered glutamate very well. These types of observations led to the removal of the food flavoring monosodium glutamate from infant foods, for fear of the remote possibility that glutamate could cause hypothalamic damage in humans as well as in rats.

LEPTIN AND THE HYPOTHALAMIC CONTROL OF APPETITE

Probably the most dramatic example of a calorie signal detected by the hypothalamus is a small protein secreted by fat cells called leptin. The discovery of this protein is a fascinating story in itself. It began in the 1950s with the identification of an abnormally fat mouse in the Jackson Laboratories of Bar Harbor, Maine, an institution famous for housing hundreds of different strains of mice used in research. This fat mouse, called an obese mouse (and possessing the ob/ob gene), was soon found to harbor some kind of mutation that caused it to eat too much and to become very fat. Fat accounts for 90% of the body weight of these chubby mice, as compared to only 20% or 30% of the body weight of a normal mouse! The unanswered question about these mice was how could a mutation in a single gene cause this severe type of obesity?

In the early 1990s, Dr. Jeffrey Friedman of Rockefeller University became determined to sort out which chromosome contained the mutation and to find the DNA region that contained the ob/ob gene. This was a real challenge. We now know that the DNA of human chromosomes contains about 22,000 DNA sequences that code for proteins (the definition of a gene). However, these 22,000 sequences amount to only about 1% of the total DNA in a cell; the rest of the DNA, interspersed among the genes on a chromosome, doesn't code for anything! Thus, a scientist cannot simply write down all the DNA sequences in an entire chromosome and expect to find a single gene among all his data; even if he could, this would be an enormous task. So how could Dr. Friedman expect to find his mysterious obesity gene?

The foundations for the kind of analysis that Dr. Friedman needed were laid down decades earlier in the lab of a famous geneticist named Thomas Hunt Morgan, who worked at Columbia University. Dr. Morgan's specialty was the genetics of a tiny fly called the fruit fly (*Drosophila melanogaster*). A lot of us now would wonder why anybody would care about the genetics of a fly, but in fact this fly was an ideal experimental animal. In cells from the salivary glands of this fly, the chromosomes lined up in multiple copies that adhered to each other so that giant chromosomes could be isolated and stained to examine the minute banding patterns on them. This gave some hope that the structure of these chromosomes could be understood. Also, flies feed on cheap concoctions made of molasses and cereal, and a scientist can breed thousands of them in short periods of time within test tubes that fit nicely into the confined spaces of even the most modest of labs. Finally, when exposed to X-rays that damage DNA, flies develop all sorts of mutations that can be passed on to their offspring.

In 1911, using these types of methods, Morgan was able to create flies with inheritable abnormalities. For example, normal flies have red-colored eyes and a gray body color, but X-rayed flies can develop white-colored eyes or a yellow body color because the genes for these traits have become abnormal. Morgan bred red-eyed, gray flies with white-eyed or yellow flies. He expected that the babies of these flies (progeny) would have random assortments of features; that is, flies could have red eyes and a gray body, red eyes and a yellow body, white eyes and a gray body, or white eyes and a yellow body. But this was not what happened at all. It turned out that fly babies were much more likely to have both red eyes and gray bodies than would be expected by chance. This, Morgan proposed, could only happen if the genes for eye color and body color were on the same chromosome and thus would mostly be inherited together. These genes were thus linked on one chromosome.

Once it was determined that two genes were linked together on one chromosome, it even became possible to find out how far from each other these genes were. This is because in the process of cell division that occurs during sperm or egg formation, the DNA of one (maternal) chromosome can be cut and then exchanged with the DNA strand of another (paternal) chromosome (see more about this in chapter 3). This will cause the two linked genes to suddenly become unlinked. The farther the two genes are apart on one chromosome, the more likely it will be that they can become unlinked by this mechanism. This type of analysis showed which fly chromosome contained the genes and even identified which bands on this chromosome were near to the genes. This pioneering method soon became applied to mice, and after a while the genes for many types of proteins were localized to specific spots on specific mouse chromosomes.

This set the stage for Dr. Friedman's work. He bred about 350 obese mice with 350 normal mice and analyzed their offspring. He found that if a baby mouse inherited the gene for obesity, he also always inherited the gene for a specific form of an enzyme called carboxypeptidase A (this is a protein produced by the pancreas that helps us digest meat). Since it was already known that the carboxypeptidase gene was on chromosome 6, this meant that the obese gene must be close to it on chromosome 6. Additional analysis of another 835 baby mice derived from more cross-breeding narrowed down the region of chromosome 6 that had to contain the obese gene. Finally, specific enzymes were used to break chromosome 6 into fragments, and the sequence of the nucleotide bases on the obese gene fragment was laboriously analyzed. All of this painstaking work took about eight years to complete.

In 1994, Friedman and coworkers were finally able to publish the structure of the gene that was responsible for the obesity of genetically obese mice. This ob/ob gene codes for a small protein secreted from fat cells that was named leptin (Greek for "slender"). As fat cells fill up with lipid, they secrete more and more of this protein, which is carried up into the arcuate nucleus through leaky capillaries. There, it influences the function of nerve cells that control feeding.[13] Thus, increases in obesity and fat cell size, at least in theory, should diminish feeding and appetite by causing a greater production of leptin.

The reason why ob/ob mice become fat is that they produce an abnormal form of leptin that fails to signal the hypothalamus that the mouse is getting too obese. After this discovery, it was also found that another obese strain of mouse, called the db/db mouse, and an obese strain of rat called the Zucker "fatty" rat, derive their own features from another problem with the leptin system: these rodents make a perfectly normal form of leptin, but they have an abnormal and unfunctional form of the receptor for leptin in their hypothalamic neurons (Fig. 1-4). The hypothalamus cannot detect that leptin is present in the blood, so these strains of rodents acquire the same features as ob/ob mice.

Do these discoveries in rodents have any application to human beings? An inability to make a normal form of leptin in humans can cause the same symptoms that are seen in rats and mice. In the 1990s, a number of patients with inherited defects in the leptin gene were identified. These rare examples of leptin deficiency appeared in several families from Pakistan and Turkey. One patient was a little girl who had a normal birth weight but who soon developed a voracious appetite when fed infant formula and became noticeably obese by her third month of life. By age five, the little girl weighed over 140 pounds, had developed asthma, and was definitely unhealthy. Doctors were able to dramatically reduce her obesity and appetite by giving her injections of leptin, so that by age nine she had a much more normal anatomy and physiology.[7]

Which nerve cells are affected by leptin, and where do they send information in the brain? The arcuate nucleus contains thousands of nerve cells, but not all of them are sensitive to leptin and not all of them regulate feeding. How are the feeding-regulating neurons to be identified?

Answering this question also required the development of more modern scientific techniques. While many nerve cells look basically alike, they differ in the types of chemical neurotransmitters they use to communicate with

Figure 1.4.   This picture shows, on the left, a mouse with an inactive leptin receptor in the hypothalamus, which causes it to overeat and become obese. A normal sized mouse is shown on the right.

other nerve cells. For example, some of the nerve cells in the lateral part of the arcuate nucleus use a small molecule called dopamine that is secreted from axonal endings (synapses) to activate other neurons (Fig. 1-5). As we will show later, these dopaminergic neurons allow the hypothalamus to control the secretion of a hormone called prolactin from the nearby pituitary gland. How can a neuroscientist identify these neurons?

Dopamine-containing neurons can be distinguished from adjacent neurons because they make an enzyme called tyrosine hydroxylase to convert the amino acid tyrosine to dopamine. If this enzyme is isolated from rat tissues, purified, and injected into a rabbit, the immune system of the rabbit will react to it as a foreign, invading protein and will synthesize antibody molecules that will specifically stick to it and mark it for destruction. In other words, the rabbit will become "allergic" to the hydroxylase enzyme and will make chemicals that bind to the enzyme.

A researcher can utilize the blood serum from an immunized rabbit to specifically stain cells that contain tyrosine hydroxylase and dopamine. The serum is diluted and poured onto a slide bearing a thin section of the

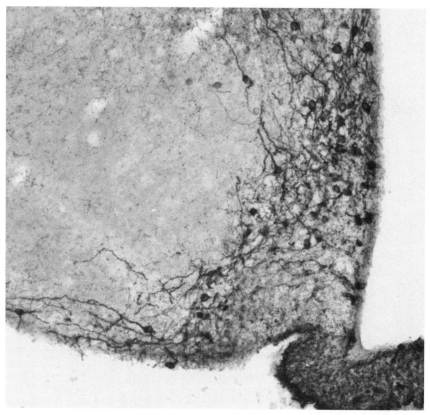

Figure 1.5. High magnification view of the arcuate nucleus of the hypothalamus, showing dopaminergic neurons stained darkly by immunocytochemistry. Cells of the nearby ventromedial nucleus (just to the left and above the arcuate nucleus) seem to create some sort of barrier around the ventromedial nucleus that is impermeable to the cells and processes of the arcuate nucleus. Thus, the ventromedial nucleus in this picture appears as a completely empty, egg-shaped structure devoid of dopaminergic processes.

hypothalamus. After washing with buffer, most of the rabbit blood proteins will flow away, leaving behind only the antibodies to tyrosine hydroxylase that have stuck to the enzyme molecules in the brain tissue. Then, using dye molecules that attach to the antibodies, a researcher can specifically stain dopaminergic neurons and leave other types of nearby neurons unstained. This procedure, called immunocytochemistry, can be used to identify many different types of neurons that utilize many different types of neurotransmitters. The whole process sounds difficult, and actually it is a bit tedious, but it has been made much easier by commercial research companies that prepare easy-to-use kits for antibody staining. These companies have gathered hundreds of different antibodies from research labs around the world for sale. These days, it is possible to stain a section of brain tissue for almost any protein one would desire, just by ordering the right chemicals.

These types of procedures have identified two populations of nerve cells within the arcuate nucleus that respond to leptin and regulate feeding. One population of cells, located close to the third ventricle, contains two different neurotransmitters called neuropeptide Y (NPY) and agouti-related peptide (this peptide was first identified in another obese strain of mice that also had peculiar yellow, or "agouti" colored, hair). These peptides are strong stimulators of feeding: when as little as five micrograms of NPY are infused into the third ventricle of a rat, that rat will eat five times more than a normal amount of food over the subsequent two hours.[22]

Other clever techniques have confirmed that leptin specifically acts upon NPY-containing neurons to depress feeding. One technique is to infuse the hypothalamus with a form of NPY that is linked to a poisonous molecule called saporin. Neurons that normally contain NPY usually release it into their surroundings and then specifically take it back up into the cell so that it is not wasted. When NPY-containing neurons are tricked into taking up NPY linked to saporin, they die but leave other adjacent neurons completely unaffected. Rats treated in this way become completely insensitive to the feeding-restraining effects of leptin because they lack NPY-containing neurons in the arcuate nucleus.[2]

These types of experiments have finally proven that nerve cells within the hypothalamus itself have a commanding influence on feeding. It now seems certain that the effects on appetite of electrolytic lesions that were first seen 60 years ago were *not* entirely caused by an interruption of nerve fibers passing through the hypothalamus but resulted from damage to hypothalamic cells themselves. What now remains to be sorted out is how these nerve cells cause their effects. One way that NPY-containing neurons stimulate feeding is to send projections (axons) to the nearby paraventricular nucleus. Neurons in this nucleus that utilize oxytocin as a neurotransmitter project down to areas of the medulla that control swallowing and chewing.[15]

Another population of arcuate neurons contains a neurotransmitter with the ungainly name of proopiomelanocortin (POMC). These neurons are strong inhibitors of feeding. The reason why GTG lesions of the arcuate nucleus result in overeating and obesity is that such lesions destroy POMC-containing neurons that restrain appetite.[12] The simple concepts of a "satiety center" and a "feeding center" in the hypothalamus must now be revised to account for the variety of different nerve cells that affect feeding in different ways.

The discovery of leptin as an appetite-restraining hormone provoked a wave of excitement within the pharmaceutical community. At that time, thousands of Americans were spending millions of dollars every year on diet books, gym memberships, and other activities in frustrating and often futile attempts to lose weight. What if all of these dollars could be spent on a leptin-like drug instead? What a potential bonanza for a drug company! In 1995, one such company, Amgen, took the risk of spending $20 million just to acquire the commercial rights to leptin and then started conducting research to see if leptin could reduce appetite in humans as well as in obese mice. Finally, in 2000, after numerous experiments in rodents and a clinical trial in humans, the results came in: most obese people already have elevated blood levels of leptin, and giving them additional leptin treatments did little to reduce their appetite and body weight. What a disappointment![3]

If leptin inhibits the hypothalamic neurons that stimulate feeding, how is it possible that any of us can become fat? Wouldn't elevations in blood leptin levels, seen in obesity, prevent the hypothalamus from causing further overeating? And why did treatment of obese people with leptin fail to reduce their weight?

The answer to this question seems to be that certain diets that are both tasty and rich in fat tend to override the inhibitory signals provided by leptin. In particular, diets rich in a specific fatty acid, palmitic acid, seem to provoke metabolic changes in hypothalamic neurons that render them resistant to the appetite-suppressing effects of leptin and also of insulin.[1] This may be one reason why diets rich in animal fats (along with carbohydrates) tend to promote obesity. In contrast, diets rich in oleic acid, which is derived from olive oil, do not seem to provoke this leptin resistance and improve the overall health and leanness of a rat. A major goal for researchers of obesity and feeding is to find a way to prevent high-calorie diets from making our hypothalami insensitive to leptin. If this could be accomplished, our bodies would automatically resist becoming obese, regardless of how much tasty food is available!

Is there any evidence that obesity in most people could result from a genetic defect like that seen in the ob/ob mouse? There are many examples of unusual obesity in people, but apparently very few of us become obese because of an abnormal leptin molecule or an abnormal receptor for leptin in the hypothalamus. Only a few families with such inherited disorders in leptin signaling have been found in the entire world. So, other reasons for obesity (superabundance of tasty foods that promote leptin resistance, inactivity, etc.) must still be sought.

All of this interest in the arcuate nucleus helped explain why arcuate lesions caused obesity, but it still failed to explain why lesions on either side of the arcuate nucleus caused a diminished appetite. Now, however, we at last know the answer: lateral hypothalamic lesions destroy other populations of nerve cells with their own specific neurotransmitters. These cells, which contain chemicals called orexin or melanin-concentrating hormone, stimulate appetite (orexin-containing cells also influence sleep and will be discussed in chapter 4). So, lesions that destroy these cells cause a long-lasting decline in appetite and body weight.

THE HYPOTHALAMUS AND
DISORDERS OF APPETITE IN HUMANS

The Hypothalamus and Dieting

All of the complicated neuronal circuits in the hypothalamus seem designed to maintain a steady body weight for each individual. By monitoring levels of nutrients and leptin in the blood, hypothalamic neurons ensure that we have enough calories on board to survive. This concept has been termed the "set point" hypothesis, in which a certain body weight is actively defended by the hypothalamus, which thus resists attempts to change it. This idea is related to how a thermostat setting maintains a comfortable temperature in a house: when it gets too cool, the heater comes on, and when it gets too hot inside, the air conditioner comes on. Similarly, the hypothalamus seems designed to increase appetite when we get too lean and decrease it when we get too fat.

This "set point" seems to work well for the majority of the population and is reasonably accurate. The problem seems to be that the hypothalamus reacts much more strongly to a loss of body fat than to a gain in body fat. This kind of makes sense: a critical loss of body fat will kill you, while a gain of body fat has deleterious effects only after quite a few years. The hypothalamus just doesn't seem to prevent obesity as well as it prevents starvation.

It could be that obese people have a genetically established "set point" that is simply higher than that of the rest of us. This might be one explanation for why we can lose drastic amounts of weight by dieting but find it so hard to keep those extra pounds off over a number of years. In a careful study of obese people put on a strict diet that simply reduced caloric intake, almost everybody was able to lose 40 pounds or so over a year. However, when these same people were followed up over several years, only about 12% of the dieters were able to keep most of this weight off, and over 40% had gained all the weight back plus a little extra![9] The only apparent predictor of successful weight loss was that the thinner people had continued to exercise moderately after finishing their diet, while the people who gained it back most easily were the ones who watched the most television. Most of us—some through bitter experiences of repeated weight loss and gain following many types of diets—are familiar with the futility of attempts at permanent weight loss through dieting alone.

The failure of diets that are simply low in calories to produce a permanent loss in weight has led to the development of many so-called fad diets that are designed in the hope of doing better. For example, the popular Atkins diet involves eating lots of meat and fats (as in bacon and steak) and avoiding carbohydrates. This diet does not aim so much at reducing the overall number of calories eaten, like other diets, but instead changes the balance of different types of nutrients within the diet. After six months, this approach does seem to allow people to initially lose about 10 pounds more than people on other types of diets; however, after a year on the diet, people on the Atkins diet were in no better shape than people on other types of diets.[18] So far, the most effective approaches toward weight loss seem to involve eating lots of foods that are filling but not too dense in calories (such as soups, fruits, and vegetables) and also regular, moderate exercise.

Surely some of the causes of the obesity of the current era stem from the ready availability of cheap, tasty foods and the lessened needs for physical activity.[20] An imbalance between caloric needs and caloric intake need not be very large to cause a gradual increase in weight. It has been estimated that people between the ages of 20 and 40 currently gain an average of about 1 to 2 pounds every year, so that many of us will be 10 to 20 pounds overweight by the time we get to middle age. To gain this weight, we need eat only 100 calories more than needed each day, which is only about 5% more calories than we need. It is likely that the hypothalamus just does not measure this small amount of calories with enough accuracy each day to allow us to easily avoid gradually putting on weight.[11]

Another possible contribution to the obesity epidemic is not even directly related to how much food we eat. Metabolic factors can also determine how efficiently fat cells retain their unwelcome cargoes. This can be easily demonstrated in obese Zucker rats that lack a functional hypothalamic receptor for leptin.

The technique to measure a rat's ability to put on weight is a simple one called pair-feeding. In pair-feeding, a fat rat is paired up with a normal, lean rat, and the fat rat is given a daily ration of powdered rat chow that exactly matches what the lean rat had eaten the previous day. After some months, both the lean and fat rats are weighed. In all cases, the genetically obese Zucker rats will put on more weight than the lean rats, even though their daily caloric intakes are the same. This is probably because the leptin-insensitive hypothalamus of the Zucker rats not only strives to increase food intake but also maximizes the efficiency of nutrient utilization to promote obesity. The precise ways the hypothalamus can do this are well known: by manipulating blood levels of hormones like insulin or cortisol, by influencing the burning of calories in a type of fat called brown fat (see chapter 2), or by simply causing less locomotor activity (walking around less often), a rat can be made to store fat much more efficiently than normal.

Some of these factors doubtless come into play in obese humans as well as in obese rats. In one clever study, small devices called accelerometers were attached to 20 lean and 20 obese people. These devices allowed the experimenters to measure the postures of each person for 24 hours and also allowed examination of energy "wasted" in fidgeting, getting up and down, and the like. The results showed that mildly obese people sat or reclined for almost two and a half hours longer each day than leaner people, and they also tended to "fidget" less. When the obese people were put on diets to lose about 10 pounds, they still persisted in lying or sitting more than lean people.[16] These types of studies show that even minor expenditures of calories that do not involve strenuous exercising can still affect how much fat we put on.

It is not even clear that people who are moderately obese actually eat significantly more than leaner people. It is difficult to accurately estimate the precise number of calories that people eat in a nonhospital setting, and many studies have found no significant differences in total daily caloric intake between otherwise normal obese people and lean people. Curiously, a study of 4500 Canadians reported that people who ate a diet relatively high in carbohydrates were less likely to be obese, perhaps partly because the carbohydrate foods that they ate included fruits and vegetables that are high in fiber.[18]

Why is it that some of us seem to be able to eat as much as we want without gaining too much weight, whereas others seem to get fat if we eat just one more potato chip each week? Genes seem to play a major role in determining which of us are at risk for obesity. One of the earliest investigators of how we inherit genes that cause obesity was a psychiatrist named Albert Stunkard, who worked at the University of Pennsylvania. During the 1980s, he studied records on several thousand identical or fraternal twins and found out that if one identical twin gained weight over his life, his brother was relatively likely to also gain weight. However, if a fraternal twin gained weight, his genetically different brother was half as likely to also gain weight than if he had been an identical twin. These types of studies suggest that about half of the variability in body weight between people is due to inherited genes.

What are the genes that tend to make us fat? As noted earlier, mutations in leptin or leptin receptors can cause obesity, but these types of mutations are extremely rare. Over the last five years or so, geneticists have identified more common variations in dozens of other genes that seem involved in routine types of obesity. One of these is the so-called fat mass and obesity-associated gene (FTO gene). About 16% of people harbor the "wrong" form of this gene, which almost doubles the risk for developing obesity. The protein coded for by this gene is more abundant in brain cells than in other cells, suggesting that a change in the function of the brain and hypothalamus must be the reason why people with this genetic trait tend to become fatter. Other more recently discovered variations in genes also seem to affect hypothalamic function. More study of these genes is needed to see if drugs can rescue us from the fate of getting fatter.[31]

For most of us, a hypothalamic lesion does not explain why we gain weight. Arranging your life so that fattening foods are not present in the house and so that you need to walk to the grocery store may be the best solutions for keeping off weight that are available at the current time. I have my own modest suggestion: buy a small two-wheeled cart and walk it to the grocery store whenever you want to buy food. Since the capacity of the cart is limited to one bag of groceries, the amount of food you could eat when you get home would be small before you would have to take another walk again. I could call this diet the "two-wheeled cart diet." Of course, I don't precisely know whether it would really work or not, but this problem doesn't seem to have stopped the authors of the other hundreds of diet books now on the market! Perhaps, as the appetite-controlling circuitry of the hypothalamus becomes better understood, it might someday be possible to develop chemicals that stimulate the hypothalamus to avoid the temptations of tasty foods and eliminate the need for all of these dieting schemes.

The Prader-Willi Syndrome

This medical condition is a relatively rare genetic disorder (1 out of 15,000 births) that involves a number of symptoms that rather resemble Fröhlich's syndrome (described earlier). The Prader-Willi (PW) syndrome was first described by a team of Swiss physicians (including Andreas Prader and Heinrich Willi) in 1956. Infants who are born with this disorder first exhibit a poor muscle tone as their main symptom, along with variable degrees of mental disability (e.g., delayed talking). Later, when they are toddlers, they develop an insatiable appetite and tend to eat any type of tasty food that is available. They crawl over to kitchen cabinets and the refrigerator and eat whole jars of peanut butter or a pound of butter all at once. Parents are often forced to lock up the doors of their kitchens and design special diets for their children to prevent them from getting very obese. As teenagers, PW children fail to show signs of sexual development, suggesting a generalized problem with the hypothalamus.

What causes this disorder? It has been difficult to answer this question, since PW syndrome is different from many other genetic syndromes. In many genetic diseases, a single-point mutation in chromosomal DNA leads to the production of an abnormal amino acid in a protein and causes the protein to function abnormally. However, in PW syndrome, a long stretch of DNA on chromosome 15 is affected (it is either completely missing from the chromosome, or else insufficient methyl groups are added to the DNA in this region). This region contains genes for at least seven proteins and also contains a region that codes for an unusual type of RNA called small nucleolar RNA (SNORD116).

To see which of these DNA segments might be responsible for PW syndrome, researchers have created genetically altered mice that lack one or all of these genes. Mice lacking the SNORD116 genes tend to eat a bit more than normal, but they never actually become obese.[5] On the other hand, mice lacking other genes within the chromosome region affected in PW syndrome (called the necdin gene or the closely related Magel2 gene, which both affect the development of neurons) don't seem to eat excessively, but they nevertheless become fatter than normal mice and also have reproductive impairments.

Several recent studies of the Magel2 gene have provided promising results. The Magel2 gene belongs to a large family of genes that were first identified in skin tumors called melanomas. The name of this gene stands for melanoma antigen gene expression, subfamily L2. When mice are genetically engineered to possess an abnormal Magel2 gene, they grow up to have some of the feeding abnormalities seen in PW syndrome, due from an abnormal production of oxytocin by the paraventricular nucleus. Also, Magel2 is produced in particularly high amounts by the arcuate and suprachiasmatic nuclei of the hypothalamus, and if this gene is abnormal, these hypothalamic nuclei don't regulate hormone release normally.[23,29] Thus, the precise causes of PW syndrome, and how they relate to hypothalamic function, seem closely linked to abnormalities in this gene. These types of studies show how useful experiments in mice have been in explaining how the hypothalamus of humans may become dysfunctional.

Anorexia Nervosa

I initially started my studies as a graduate student by trying to understand anorexia nervosa. I had hoped that by learning about the hunger-regulating circuits of the hypothalamus, I might better understand anorexia and perhaps could even perform research that might provide clues about it. These ideas led me to a lab (Dr. Roger Gorski's lab) in which I could study feeding behavior in rats. I was guided in these efforts by Dr. Dwight Nance, a postdoctoral researcher in Gorski's lab who became a good friend of mine and who had lots of strong and colorful opinions about how the hypothalamus worked. He helped me to come up with a number of experiments that analyzed how the female sex hormone, estrogen, affects how much a rat will eat. As you will see, these experiments actually had some relevance to anorexia nervosa.

What is anorexia? Anorexia nervosa is a disorder in which patients exhibit a sustained decrease in food intake that leads to very severe emaciation. It has been known for hundreds of years. Anorexia is much more common than the relatively rare Prader-Willi syndrome; as many as 1 out of 500 people may experience one or more episodes of this disorder during their lifetime. Anorexics may starve themselves down to 60 or 70 pounds, which is dangerous for health and which accounts for the fact that the mortality rate for anorexics is higher than for any other psychological disorder. Some anorexics accomplish this by obsessively monitoring each bite of food they eat and counting all their calories. Other anorexics may take excessive amounts of laxatives or undergo self-induced vomiting to lose weight. Anorexia nervosa is unusual in that it shows a dramatic sex difference in incidence: at least 90% of anorexics are female, and the disorder usually begins during adolescence or soon afterward. There have been countless attempts to explain the reasons why people become anorexic.

For the last 50 years, the predominant view of anorexia has been that it is a psychological illness originating from some sort of stress. An unhealthy home environment, fears of sexual development, perfectionistic attitudes toward achievements or body size, and so forth have all been hypothesized to contribute to anorexia. I have gotten acquainted with several women who have had episodes of anorexia. Some of them seem to have recovered a relatively normal body weight and have abandoned the preoccupations with body weight and dieting that they held onto so strongly years before. Others still obsessively diet and exercise to keep their body weights down to low levels. Aside from these characteristics, I have not been able to discern any psychological abnormalities in these few subjects, although I am not a professional psychologist or psychiatrist.

More recently, the viewpoint that anorexia is purely a psychological disorder has been changing. Some of the abnormal psychological attitudes towards food, eating, and body size seen in anorexics can also be provoked in ordinary people by prolonged fasting: starving people also have abnormal preoccupations with food and body size.[27] Starvation can also bring about a loss of the menstrual cycle that is seen in anorexia, since decreased amounts of body fat and decreased levels of leptin prevent the hypothalamus from stimulating the release of reproductive hormones from the pituitary gland (see chapter 5). Finally, it has been recognized that there is a strong genetic component to anorexia. For example, identical twin girls are substantially more likely to both develop anorexia than are fraternal twin girls. Since identical twins share exactly the same genes, unlike fraternal twins, these findings suggest that there is some biological factor that also contributes to the risk of developing anorexia. But what is this factor?

Numerous studies of genes and molecules that could affect the hypothalamus and potentially trigger anorexia have been carried out. Many of these studies have examined genetic variations in neurotransmitters and their receptors to see if anorexics have a mutation that could explain their symptoms. Genes for neurotransmitters like dopamine, serotonin, and norepinephrine, for example, have all been examined. Some of these studies initially looked like they had identified a gene that promotes anorexia, but later these studies were not confirmed. Still, genetic studies continue to be vigorously pursued. It may be that many genes interact to produce the symptoms of anorexia, which would make a genetic analysis much more complicated.[24]

Perhaps the most curious thing about anorexia is that most sufferers are female. This may simply reflect the pressures of society that dictate a slim appearance for women. However, there are legions of fat people in the population who are strongly motivated to lose weight and are often on diets, but few manage to drastically reduce their body weights permanently.[9] Something stronger than societal pressure must be enabling anorexics to chronically maintain such a low body weight.

One possible explanation for the weight loss of anorexics and the sex difference in incidence is an influence upon the hypothalamus by the female sex hormone, estrogen. In rats, estrogen can reduce food intake by an action upon hypothalamic NPY-containing neurons.[22] Similar effects in humans are probably also explainable by an action of estrogen upon the hypothalamus. Human beings exhibit changes in food intake over the menstrual cycle, and during pregnancy, when effects of estrogen are blocked by another hormone called progesterone, there is a substantial increase in appetite. Since anorexia mainly affects young girls at or around the time of puberty, I thought that perhaps an abnormal response of the hypothalamus to rising levels of estrogen at puberty could contribute to the syndrome.

My idea was a pretty obvious one, but I was nevertheless able to publish it as a small speculative paper when I was just a graduate student, which made me very happy. For the first time, I became a published scientist! Not long after my first paper was published, I received postcards in the mail from lots of different scientists asking for reprints of my paper. It was delightful for me to know that anyone else would be interested, and, as an added benefit, I could collect stamps from a variety of different countries from the postcards. I gave these to a lovely young lady named Paula Spesock who collected stamps and lived in the same graduate dorm where I also had a room. Eventually, I was lucky enough to marry her.

My interest in anorexia had led to a speculative paper. However, as they say, talk is cheap: it is easy to come up with a plausible idea for a disorder, but when it comes to actually providing real evidence for it one way or another, that is another story. How could I prove that an abnormal response to estrogen might take place in anorexia? Clinical studies of what might cause anorexia require the participation of physicians and psychiatrists, large amounts of actual work, and research money to fund all of these efforts. Thus, it was difficult to translate my idle speculations into real data.

To try to make some progress on anorexia, about 15 years ago I got in touch with Dr. Johannes Hebebrand, a scientist from Germany who was organizing a sophisticated effort to find the genes responsible for extremes in body weight. I read about his efforts in journals, sent him a letter, and found out that he would shortly be visiting relatives in the Washington, D.C., area who lived in nearby Georgetown! So, on the appropriate day, I took a crosstown bus that shuttles between Howard University and Georgetown to meet Dr. Hebebrand and see if he might be willing to look into the genetics of anorexia nervosa. He was gratifyingly interested, and I seized on the opportunity to propose that he look at his genetic samples from anorexic patients to try to see if they might have an abnormality in the estrogen receptor. He agreed and asked me to suggest how to proceed. Here, however, I led him astray: I proposed that he look at the beta form of the estrogen receptor, which at the time was considered to be the protein that mediated the effects of estrogen on appetite. Dr. Hebebrand did complete the study but came up with disappointingly inconclusive results. Now I know one of the reasons for this disappointment: we now know that another type of estrogen receptor protein, the alpha type, is the one possessed by appetite-restraining hypothalamic neurons that respond to estrogen. I had pointed Dr. Hebebrand in the wrong direction at the wrong protein, so real proof about my ideas on anorexia remained elusive.

One nongenetic type of proof that estrogen could be involved in producing the symptoms of anorexia may stem from actual clinical studies of a special population of patients with a relatively rare disorder called Turner's syndrome. In Turner's syndrome, an X chromosome is missing from all the cells of the body. This genetic abnormality somehow prevents the development of ovaries that produce estrogen, so that a young girl with Turner's syndrome will not show physical signs of sexual maturation. Not infrequently, girls with Turner's syndrome are treated with estrogen to promote sexual development. In 21 such patients, episodes of anorexia nervosa have been reported to follow this sudden treatment with estrogen.[34]

Why would these patients respond so abnormally to estrogen? It could simply be that the sudden introduction of estrogen into the body, so different from the gradual rise of estrogen that takes place over puberty, could itself provoke an abnormal response. Another possibility may stem from the other endocrine abnormalities known to be present in Turner's syndrome. In addition to having no ovaries, Turner's syndrome patients often have very low levels of thyroid hormone (thyroxine). I found, some years ago, that thyroxine can potently block anorexic effects of estrogen in rats, probably by interfering with the action of estrogen receptors on cells.[33] If Turner's syndrome patients lack a hormone that normally blocks the anorexic effects of estrogen, this may be part of the explanation for why Turner's syndrome patients respond to estrogen by developing a bout of anorexia nervosa.

The findings of anorexia in patients who had been treated with estrogen to correct symptoms of Turner's syndrome represent a relatively rare and anecdotal type of evidence that is vulnerable to many criticisms. However, finally, the hypothesis that estrogen may somehow be involved in the alteration of food intake in anorexia has been supported by a report from a group of scientists in France, headed by Dr. Nicholas Ramoz. In this report, DNA samples from 700 families were acquired and laboriously analyzed for variations in the gene that codes for the type of estrogen receptor that causes decreased food intake. This study showed that women with an abnormal form of this alpha type of estrogen receptor are more likely to develop anorexia.[30] If this recent finding is confirmed, it might finally be a piece of solid evidence that my naive idea has some merit.

The findings of Dr. Ramoz and his colleagues are very interesting, but there are some difficulties with his report. He discovered that so-called single nucleotide polymorphisms in the DNA that codes for the estrogen receptor are much more frequent in women suffering from anorexia nervosa. However, these variations in the DNA for the receptor are found in surprising regions of the gene: they are found in introns. What are introns?

In the cell nucleus, each time a gene is "read" by chromosomal enzymes, a long string of messenger RNA (mRNA) is produced that duplicates the sequence of nucleotides on the DNA of the chromosome. Later, this chain of messenger RNA is sent into the cell cytoplasm, where it is once again "read" by ribosomes that use the information on the mRNA to assemble individual amino acids into a long chain that becomes a protein. In this case, the protein is the receptor protein that binds estrogen and which is responsible for all the biological effects of estrogen upon the cell.

The problem with this simple story is that not all portions of the original mRNA molecule manage to reach the cytoplasm. Some segments, termed introns, attract the attention of a tiny nuclear machine called a spliceosome. This machine settles down onto an intron segment, forces it into a loop, and then snips this loop completely out of the mRNA molecule, at the same time rejoining the cut ends of the mRNA on each side of the loop. The final mRNA molecule that reaches the cytoplasm is thus completely free of introns and is composed solely of exons that code for a protein.

If a mutation in the DNA for the estrogen receptor is found in an intron, it wouldn't cause any change in the final structure of the estrogen receptor, because mRNA formed from introns never reaches the cytoplasm and is never translated into a protein at all. So how could the findings of Dr. Ramoz have any significance for anorexia nervosa?

The answer is that abnormalities in introns can confuse the spliceosome machinery and either cause it to fail to remove introns, cause it to remove too much from the mRNA, or else alter how many mRNAs are produced per hour. All of these things can affect the function of the final estrogen receptor protein.[26] These details illustrate some of the many difficulties in obtaining evidence about a disease through genetic analysis.

Even if so, so what? If estrogen or estrogen receptor abnormalities might be some of the factors that could worsen anorexia, how might this lead to any form of treatment? It is no good to learn a cause of a disorder if it doesn't allow a physician to help her patients. Possibly administration of some types of progesterone, which block the anorexic effects of estrogen, could be of value. Any type of treatment, however, must also be proposed with caution: anorexic girls often defend their lifestyle and feel that they should have the freedom to live as they want. As in many psychological disorders, attempts to "normalize" the life of a patient should only be carried out with the cooperation of the patient, as much as is practicable. Anorexia nervosa is only one of many behavioral syndromes that pose therapeutic challenges to physicians. How much coercive treatment is warranted in medicine, even if the treatment is for a life-threatening disorder? Should anorexics be forced to gain weight to reduce the risk of death? Should drug addicts be forced to undergo therapy to reduce their dependence on drugs and improve their health? These types of questions have long been problematic for psychiatrists and are not likely to be resolved anytime soon.

Gastric Bypass Surgeries and the Hypothalamic Control of Feeding

One increasingly common approach for treating extreme obesity is a surgical procedure called a gastric bypass. In this procedure, the stomach is divided into a large sac that is stapled shut and left in place, and a smaller sac that remains connected to the esophagus. This smaller stomach sac is then attached directly to the intestines. The small size of the stomach prevents the ingestion of lots of food at one time, and the more rapid passage of food through the digestive tract decreases the absorption of nutrients. These factors alone probably account for much of the success of this procedure, which allows patients to lose as much as 150 pounds and keep this weight off for long periods.

The physical size of the stomach, however, is probably not the only factor in the success of this procedure. As it turns out, epithelial cells lining the stomach release a protein hormone called ghrelin into the bloodstream during meals. When the ghrelin reaches the hypothalamus, it stimulates NPY-containing neurons to increase appetite. The decreased stomach activity in gastric bypass patients causes a lower secretion of ghrelin, which also contributes to the decreased appetite of these patients.

Could more moderate obesity in most people be treated by blocking the effects of this appetite-stimulating hormone upon the hypothalamus? This kind of approach has been studied in mice by a number of drug companies. Unfortunately, up until now, ghrelin-blocking experiments have not produced dramatic effects upon body weight in mice.[17] Somehow, the hypothalamus compensates for reduced actions of ghrelin, and a normal appetite seems to be present even in the absence of ghrelin. These types of studies show just how difficult it is to override the natural signals that control body weight and hypothalamic activity.

Bardet-Biedl Syndrome

One of the puzzles concerning the function of leptin-sensitive neurons and the control of obesity has involved a relatively rare inherited disorder called the Bardet-Biedl syndrome, which affects about 1 in 100,000 people in the general population. This disorder was first identified in the 1920s by two different physicians, Georges Bardet in France and Arthur Biedl in Hungary, who worked independently of each other but nonetheless described essentially the same medical condition. Children with this disheartening disorder may often be born with an extra finger or two and then start to show additional symptoms (poor growth, delayed walking) at about 3 years of age. By about 8 years of age, children begin to have problems seeing well during the nighttime and then progress to severe visual impairment or even blindness by age 15. About two-thirds of Bardet-Biedl children have learning difficulties during their early school years; most of these children develop obesity or morbid (extreme) obesity, and many of them develop kidney abnormalities as they age. Boys with this syndrome have poorly developed testicles or testicles that have not descended during development into the scrotum; girls have some reproductive development but in later years frequently have very irregular menstrual cycles.

Some sort of hypothalamic dysfunction would appear to explain the obesity and reproductive problems of these patients, but what about the kidney and eye problems, and how could the mutation of one gene cause all of these symptoms?

Recent studies have provided the answers to this puzzle. Mutations in Bardet-Biedl patients all involve proteins that control the function of sensory cilia. What are sensory cilia?

It has been known for decades that certain cells in the body possess a single, nonmovable, hairlike projection from the cell membrane called a sensory cilium. For example, some cells forming the tubules of kidneys have this peculiar structure. The function of these single cilia was not known until they became connected with a serious kidney disease called polycystic kidney disease. In polycystic kidneys, the small kidney tubules enlarge into enormous, fluid-filled cysts that don't function very well. We now know that this is due to a disordered function of the sensory cilia of these cells. Normally, these cilia are buffeted back and forth by the passage of urine flowing through the kidney tubules. This ciliary movement informs each cell that urine is being formed normally and that all systems are working properly.

If a protein in these sensory cilia is abnormal, the cilia no longer function as sensors, and the kidney cells react as if urine is no longer being formed. In desperation, they multiply all out of control and form larger hollow structures in an erroneous effort to correct this situation. While there is still no cure for this disorder, at least we now know what causes it.

Since sensory cilia are important for kidney function, this would explain the renal symptoms of Bardet-Biedl patients. Also, retinal photoreceptor cells possess sensory structures that are derived from cilia, so a ciliary abnormality could also produce retinal problems as well. But what explains the problems with the hypothalamus?

It turns out that receptors for leptin seem to be clustered on the cell membrane of a sensory cilium that is present on hypothalamic neurons. If these receptors are not incorporated into ciliary membranes, or if the cilium does not function, the neuron will no longer respond to leptin. This causes obesity and also sterility, since leptin-sensitive neurons regulate the synthesis of gonadotropins (sexual hormones) by the pituitary gland.[26]

The discovery of the cause of this relatively rare disease is leading to a "paradigm change" in neuroscience. What do I mean by this? A paradigm change is a revolution in the standard way of thinking about a subject. In traditional neuroscience, neurons were considered to talk to each other in a well-known way; that is, a nerve ending would sense something, a wave of electricity would be carried back to the cell body of a neuron via a dendrite, and the neuron would pass on this information to another neuron by sending out an axon that would synapse upon the second neuron. Now we know that most neurons in the brain also possess sensory cilia and are constantly sampling their environment and gaining information in a way that *does not require synapses!* This flies in the face of conventional wisdom and will force a reevaluation of how scientists think about brain function. The functions of synapses and axons, of course, cannot be disregarded, but now they must be put in the context of other (ciliary) ways of gathering information by neurons.

LEPTIN AND THE MYSTERY OF PUBERTY

The control of appetite by the hypothalamus may not be the only function affected by leptin. A major life-changing event—puberty and the maturation of a person from a child to a reproductively competent adult—now also appears to be affected by leptin.

A commonplace event in all of our lives is the relatively sudden rise in blood levels of sex hormones, the maturation of the testes and ovaries, and the bodily changes that take place during puberty. We all know the consequences of puberty, but the actual causes of puberty still remain somewhat of a scientific mystery. As will be discussed in chapter 5, it is well known that the hypothalamus governs the release of reproductive hormones from the pituitary gland that cause the maturation of our sexual organs. But what causes the hypothalamus to undergo this sudden change in function during the teenage years?

In the 1970s, Rose Frisch at Harvard was one of the first to propose a linkage between nutrition, body weight, and puberty. She noted that historical medical records taken over the 19th and 20th centuries showed a steady change in puberty in young girls in the United States and in other countries. During the 19th century, the age at which a girl first started to experience monthly bleeding—an event called menarche—amounted to about 14 years of age. However, by the 1960s, the age at first menstruation had dropped to about 12 and a half and seemed to still be decreasing, though at a lower rate per decade. Frisch hypothesized that puberty only occurs when a critical level of body fat had developed. Better nutrition during the 20th century allowed this level of body fat to be attained at an earlier age, leading to an advance in the timing of puberty.[14] This general type of thinking makes biological sense: for any animal, it is certainly a bad idea to begin reproducing when there is not enough food around to build up the body's fat stores.

Frisch's ideas were supported by observations of young athletes. For example, young female ballet dancers typically exercise much more than other girls in pursuit of their future profession, and they often maintain a much leaner physique. Commonly, menarche in these girls takes place almost a year later than in other girls.[6] But how does the hypothalamus "know" that the nutritional state of the body is inadequate to permit reproduction, and what signals cause a normal puberty?

This puzzle was one of the first things I tried to study when I first was hired as an entry-level assistant professor at Howard University and got my own lab and some funding from the government. My plan was to study the effects of nutrition and hormones on puberty in rats. How could I determine when a rat experienced puberty?

This problem is actually much easier to solve than you might imagine. It turns out that baby female rats have a thin membrane of skin that covers the vaginal orifice, much like the somewhat deeper membrane called the hymen that is present in immature girls. This membrane, in rats, is very sensitive to female sex hormones. When blood levels of estrogen rise, the membrane gradually thins and ruptures in an event called vaginal opening. This event is easy to detect by simply turning a baby rat upside down and inspecting her genital area. Vaginal opening in rats takes place about 35 days after birth; rats, obviously, develop into adults much more rapidly than humans. A similar process occurs in baby male rats. In immature males, the soft tip of the penis cannot be extended past a sheltering hood of skin called the prepuce (or foreskin, in humans). But, as puberty progresses, the prepuce covering the penis loosens and expands in an event called preputial opening, which is very like vaginal opening in females.

After vaginal opening, it is also easy to confirm that teenage female rats are ovulating and experiencing cyclic variations in reproductive hormones. This can be done by withdrawing a small sample of cells from the vagina with a fluid-filled eyedropper. Once the cells are placed on a slide, it is simple to examine their shapes and sizes under the microscope. During ovulation, the vaginal cells of a rat respond to estrogen by becoming large and flat; after ovulation, smaller, rounder cells dominate the vaginal interior. By this method it is possible to confirm that rats ovulate every four days, much more frequently than the 28-day cycle between ovulations of egg cells that occurs in humans. They also are much more fertile than humans. Pregnancy in rats last about 21 days; another 21 days is required to nurse the baby rats and wean them. Because a mother rat can raise a litter of 10 to 12 or so baby rats within 42 days, they can often go on to reproduce again! Theoretically, a single female rat could produce seven litters each year, almost a hundred offspring! This prodigious ability of rodents and their relatives to reproduce can pose a serious problem when they are introduced into foreign countries where they have no natural predators or reproductive competitors. Australia, for example, experienced massive overgrowth of rodent and rabbit populations when they were first introduced into the isolated island by shipping and other human activities. Rigorous campaigns of mass poisoning or introduction of rabbit diseases proved necessary to combat the unchecked reproductive abilities of these animals.

In my lab, however, the reproduction of rats occurred at a much more modest rate and allowed me to study how nutrition could affect puberty. One way of approaching this question is to simply redistribute newborn baby rats between rat mothers. For example, two litters of 10 rat pups each could be redistributed to yield one litter of 7 pups and another litter of 13 pups. The mother rats seemed to have no objections to these schemes, but each rat had a limited ability to produce milk, so that the small litters became better nourished than the large litters. As you might expect, baby rats in the small litters put on weight more rapidly and became roly-poly sooner than the rats in the large litters. The better-nourished rats also had an earlier age at vaginal opening (perhaps two or so days earlier). I tried to see if the age at puberty was related to various hormone levels by taking blood samples from the young rats. For example, since hormones like insulin and prolactin seem to respond to different nutritional states, I measured blood levels of these molecules. I also tried to see if injecting small amounts of insulin into immature rats would advance the timing of puberty.

All of these efforts, alas, didn't yield any convincing results and amounted to some of my first "failed" experiments. Although I didn't realize it at the time, my efforts were bound to fail, because the existence of the body-weight-related hormone, leptin, was completely unknown when I started my study. Thus, even the best-intentioned plans come to naught in the presence of incomplete knowledge. This is a common risk in science; most of my colleagues will agree that successful, publishable experiments amount to less than half of all the experiments we try to do, in spite of all of our best efforts.

Since these early days of the study of puberty, more recent experiments with leptin have confirmed that the presence of leptin in the blood is necessary for a normal puberty in rats, and that leptin levels steadily rise throughout puberty. Administration of leptin to rare patients with an insufficiency of normal leptin will also restore a more normal reproductive ability, as it does in rats.[7] Although blood levels of leptin do increase throughout puberty, it is less clear that leptin alone is the major signal for puberty in most humans. Another complicating factor is that blood contains a protein that binds leptin and inhibits its action. Blood levels of this protein also vary during the process of puberty and complicate analysis of the role of leptin in human puberty.[6] Clinical data also show that young girls with a deficiency of insulin (type I diabetes) also have a delay in puberty.[25] It now seems likely that both leptin and insulin not only regulate the control of food intake by the hypothalamus but also regulate the process of puberty. So, while the discovery of leptin has certainly shed light upon the connection between the hypothalamus, nutrition, and body weight in reproduction, it is also clear that the last word has not been written on the subject.

PRECOCIOUS PUBERTY

Normal puberty in most humans occurs within a relatively narrow range of ages, from about 11 to 12 and a half years of age. In a few of us, however, puberty can begin as early as age 6 or 7. This so-called precocious puberty occurs rather rarely, in about 0.2% of girls and in less than 0.05% of boys, but since puberty takes place in everybody, these small percentages of immature humans nevertheless translate into consequential numbers of people in the United States. The causes of this very early puberty are not all understood, but often they seem related to tumors of the hypothalamus or to mutations in a number of proteins that are found in the hypothalamus.[28] All of this information shows that a monitoring of body weight by the hypothalamus is not only important for the control of appetite but also affects other physiological systems like the reproductive system.

The Hypothalamus and the Control of Thirst and Body Temperature

THE HYPOTHALAMUS AND THE CONTROL OF THIRST

𝒯hirst is probably the most intense and compelling drive that humans can experience. Extreme states of thirst have been well documented by hapless survivors of treks through deserts or shipwrecks that left them stranded in lifeboats without sufficient food or water. Thirst begins to be a terrible ordeal within two or three days without water; after a week or more, people become so desperate for water that they give in to the temptation of drinking seawater or trying their own urine to moisten their lips and slake their thirst.[21] Survivors of shipwrecks have been known to survive for weeks without food, utilizing all their body fat and some of the molecules of their muscles to provide metabolic fuel; a lack of water, however, causes terrible suffering and often death within 10 days.

We now know that the hypothalamus seems to be a critical brain region that responds to water deprivation to induce thirst. But, how does the hypothalamus "know" that there is insufficient water in the body? Which neurons cause the sensation of thirst? Thirst is not simply having a dry mouth but is a much more fundamental sensation caused by an insufficiency of water throughout the body. How can the physiology of thirst be studied?

Once again, lab rats have been the dominant thirst quenchers in the research of this behavior. It is easy to weigh their drinking bottles at intervals after some sort of a thirst challenge to see how thirsty they are.

Lab rats can also tell us a lot about how a liquid solution tastes. They can't, of course, literally tell us if a solution tastes nasty, but their behavior and reactions to fluids can tell us the same thing. A common test for how tasty a fluid is involves hanging two different bottles that contain two different solutions on the cage of a rat. This test is called the two-bottle preference test. If a rat needs to replenish the salt in its body, it will prefer drinking a salty solution, even if the salt is so dilute that it can hardly be detected. The more salt a rat needs, the more it will drink of the salty solution. The same test can be used to measure the taste acuity of a rat. For example, some years ago I was able to find out that rats can taste a very dilute solution of glucose (only a 1% solution). I tasted this solution first before I presented it to rats and really couldn't distinguish it from distilled water. The rats definitely could, though, and drank twice as much of the glucose as distilled water.[24]

Angiotensin II and Thirst

So, what causes thirst for ordinary water? A dominant physiological signal that is activated during water deprivation does not originate in the brain at all but is produced by the kidneys. If water and sodium are lost from the body (imagine taking a hike in Death Valley during August and sweating out lots of water and salt to cool your body), specialized cells in the kidneys are activated. Two types of cells are involved.

One type of cell, found in a specialized part of a kidney tubule called the macula densa, is constantly exposed to urine flowing through the tubule to reach the ureter and eventually the urinary bladder. These cells continually monitor how much sodium, chloride, and water are present in the urine. If these molecules become depleted from the body, these cells respond to this emergency by secreting a small molecule called ATP into their environment.

The ATP secreted by macula cells binds to receptors on nearby cells that are called juxtaglomerular cells. These are peculiar smooth muscle cells that wrap around arterioles in specific portions of the kidney. When they bind ATP, these cells secrete an enzyme called renin into the bloodstream. Renin, in turn, attacks a circulating protein called angiotensinogen that is manufactured in the liver and secreted into the blood. Renin removes a short stretch of amino acids from angiotensinogen and converts it into a smaller peptide called angiotensin I. This peptide is still an inactive molecule; it has to be cleaved still further by another enzyme called ACE (present in the lungs) that converts it into angiotensin II.

After these effects of enzymes are completed, a highly active peptide hormone called angiotensin II is formed. AII has dramatic effects upon water and sodium conservation by the body. Some of these effects of AII are (1) a constriction of blood vessels, which increases blood pressure and prevents you from fainting if you are dehydrated; (2) a release of another hormone, aldosterone, from the adrenal gland, which causes the kidneys to conserve body levels of sodium; and (3) a stimulation of thirst.

The effects of AII on blood pressure can potentially be dangerous. High blood pressure (hypertension) can damage the brain and heart and is a serious health risk. The precise causes of hypertension are still not well understood in spite of decades of research on this subject. However, now that the pathway that generates AII has been clarified, it is possible to create drugs that interfere with the hypertensive effects of AII. One class of drugs, called ACE inhibitors, block the ability of ACE to create AII from AI. These widely used drugs offer a simple way to reduce blood pressure and have relatively few side effects.

As regards the thirst-stimulating abilities of AII, to me, at least, it is not at all obvious that a hormone produced by the kidney might affect the regulation of drinking by the brain. How did anyone come to think up this connection in the first place? It all started in the lab of Robert Tigerstedt, a professor of physiology at the Karolinska Institute in Sweden (this same place is where Nobel prizes are decided upon today). In 1897, Dr. Tigerstedt was scheduled to chair a scientific session in Moscow and wanted to perform some novel experiment to discuss at the meeting. He decided to crush fresh rabbit kidneys, make water-soluble extracts of them, inject them into rabbits, and observe what happened. The results were that these crude kidney extracts caused a rise in blood pressure. After presenting these results, Dr. Tigerstedt lost interest, and the rest of the world lost interest along with him. His results were forgotten for 40 years! It was only in the late 1930s that the subject was approached again and the molecule in his kidney extract (renin) was identified and found to generate angiotensin I when mixed with fresh horse serum.[19]

The ability of AII to stimulate thirst as well as to increase blood pressure was not even appreciated until the late 1950s. At this time, researchers were investigating the effects of removing the kidneys from rats. Such nephrectomized rats could survive for at least a few days, but they were peculiar because they never seemed to drink water and thus rapidly lost weight. However, when they were given renin, which restored circulating levels of AII to normal, the rats began to drink water again. Clearly, AII must have been acting somewhere in the brain to cause this.[5]

However, where in the brain does AII act to produce thirst? More importantly, how can a large molecule like AII cross the blood-brain barrier and even enter the brain at all? You will recall that the brain detects changes in the overall fat and calorie content of the body by responding to leptin. Leptin can enter the hypothalamus by crossing through leaky capillaries in the median eminence and arcuate nucleus. Are these same capillaries utilized by AII?

As it turns out, the arcuate area of the hypothalamus is not the only region with leaky capillaries. A structure at the extreme anterior border of the hypothalamus, called the organum vasculosum of the lamina terminalis (OVLT), also has leaky capillaries. It is in this anterior region of the hypothalamus (median preoptic nucleus) that AII exerts most of its effects.[5]

The behavioral effects of AII have been studied by using the technique of intraventricular cannulation. In this technique, a tiny hole is drilled through the skull of a rat (under anesthesia) and a very thin, hollow metal tube (a cannula smaller than the fine pins used to attach shirts to shirt cardboards) is lowered into the brain so that it pokes into the third ventricle. The cannula is cemented into place using dental cement and plugged with a tiny wire. Later, when the rat wakes up and is walking around its cage, flexible plastic tubing can be attached to the cannula to allow tiny amounts of fluid (1/100 of a milliliter) to be slowly infused into the brain.

When small amounts of sterile saline are infused via this method, the rat will typically not even notice it and will continue to eat food or wander around its cage. However, when saline containing tiny amounts of AII (100 nanograms, or 1/10,000 of a gram!) are infused into the hypothalamus, even a hungry rat will immediately stop eating and run over to its water bottle. AII can cause a rat to drink as much as 15 mL of water over 15 minutes; normally, a rat will not drink much more than this during an entire day! Thus, AII powerfully stimulates drinking by activating nerve cells in the anterior hypothalamus. The human hypothalamus has receptors for AII in just the same region as that of the rat, so it seems very likely that the thirst we experience during dehydration is caused by AII, just as in the rat.

If we accept the proposal that the extreme anterior portion of the hypothalamus functions as a sensor for dehydration, how does activation of this area result in the sensation of thirst and in drinking behavior?

It has been established that nerve cells sensitive to AII send projections to more posterior structures like the paraventricular nucleus (PVN) of the hypothalamus (Fig. 2-1).[4] This nucleus, in turn, projects to many other brain areas. Some of these PVN projections go to the medulla of the brain, where clusters of nerve cells that innervate the esophagus and tongue are located. Activation of these nerves is necessary for the swallowing movements performed during drinking and eating. This pathway is the means by which the hypothalamus controls water intake and food intake.

What about the sensation of thirst itself? Information about this topic is less clear, but it seems likely that a portion of the cerebral cortex located just above the hypothalamus (cingulate cortex) receives input from the hypothalamus and is important for this subjective sensation during dehydration.[13]

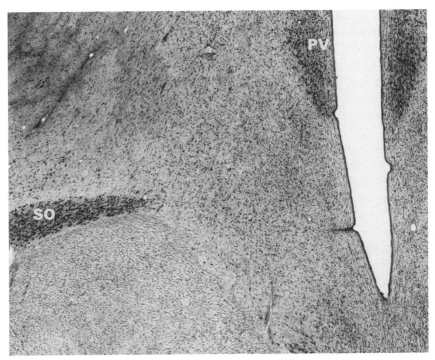

Figure 2.1. Cross section of a dog hypothalamus, showing the dark-staining collections of large neurons in the posterior part of the supraoptic (SO) nucleus and in the paraventricular (PV) nucleus. The large, ovoid optic tract is visible at the bottom of the brain.

Vasopressin and Dehydration

The anterior area of the hypothalamus is not the only region to react to a deficit in body water. When we get dehydrated, more posteriorly located neurons of the supraoptic nucleus (SON) react by sending a pulse of electricity down their axons, which terminate in the posterior pituitary gland (Fig. 2-1). Large swellings at the ends of these axons contain a powerful hormone called vasopressin (also called antidiuretic hormone, or ADH). When activated, these swellings, called Herring bodies, release ADH into the bloodstream.

ADH is important for the control of body water for a number of reasons. First, ADH can constrict blood vessels and raise blood pressure to keep us from fainting, like AII. More importantly, ADH acts upon kidney collecting tubules to prevent them from losing water into the urine. This prevents further dehydration and conserves body water.

Without ADH, the kidneys are unable to produce a concentrated urine. The consequences of this are dramatically illustrated in a genetically abnormal type of rat called the Brattleboro rat, which was discovered in a research rat colony in the 1960s. These poor rats have an impaired ability to make ADH, and consequently their kidneys are constantly producing a dilute urine. Such rats have been observed to drink as much as 70 mL of water over an eight-hour period, almost 10 times more than the water consumed by a normal rat. This excessive drinking makes up for all the water lost in the urine.

A similar condition, called diabetes insipidus, exists in humans. Diabetes insipidus was first properly understood by Alfred Frank, a German physician who examined a patient with a puzzling variety of symptoms in 1912. His patient had been accidentally shot in the head in a hunting accident. This poor man recovered pretty well from his misfortune, but he had one nagging problem: he was constantly thirsty, drank huge amounts of water, and urinated a proportionately large amount of urine every day. When Frank took X-rays of his head, he was astonished to see that the bullet had lodged in the sella turcica, a cavity in the sphenoid bone of the skull that contains the pituitary gland. Apparently the bullet had obliterated the posterior pituitary gland and eliminated the nerve endings that contained ADH. This caused his patient to urinate excessively.

Nowadays, most cases of diabetes insipidus are caused by tumors of the pituitary gland that encroach upon the hypothalamus and damage nerve cells in the supraoptic nucleus. Patients with this condition lack ADH and can produce 8 to 20 quarts of urine over 24 hours; they must drink a correspondingly great amount of water to prevent dehydration.

Fortunately, diabetes insipidus can be treated relatively easily. ADH is a small protein, composed of only nine amino acids. Since it is so small, it can be dissolved in saline used in a nasal spray. A synthetic version of ADH, which works well on the kidney but does not cause vasoconstriction, can be sprayed into the nasal cavity of a patient with diabetes insipidus every eight hours or so. The small molecules of ADH are readily absorbed into the bloodstream through the mucous membrane lining the nose and can thus prevent the symptoms of diabetes insipidus.

How do ADH-producing supraoptic neurons "know" that a deficit in water is present in the body so that they can react by releasing ADH into the bloodstream? These neurons actually have a means of sensing this directly. When our bodies lose water, the blood circulating within capillaries becomes thicker, saltier, and more viscous. This salty blood draws water out of cells by causing an osmotic gradient of molecules and a transfer of water through the cell membrane. This is analogous to cooking a hot dog in very salty water: such a hot dog will shrink instead of swelling, which normally happens when a hot dog is cooked in pure water. The shrinkage of supraoptic neurons causes them to fire a pulse of electricity down their axons.

Once again, however, it is reasonable to ask how the neurons "know" that they have shrunk. This question has only been answered recently, thanks to the discovery of a large family of proteins that function as sensors for sensory neurons. Proteins in this family are called transient receptor potential vanilloid proteins (TRPVs). They have an interesting history.

TRPVs were first isolated by researchers interested in learning how sensory neurons react to heat. These researchers knew that capsaicin, a flavoring compound present in hot chili peppers, could trick sensory nerve endings in the mouth to react as if they were being exposed to true heat (we all have had this experience when eating spicy foods!). It turned out that this chili pepper compound, capsaicin, was binding to a sensory protein in the cell membrane of nerve cells that normally opened up in response to an elevation in the temperature within the mouth. This allowed calcium ions to enter nerve endings and caused them to generate an electrical signal. Other plant compounds like menthol activate cold-sensitive nerve endings. By isolating the proteins that bound capsaicin and menthol, researchers were able to identify heat- and cold-sensitive proteins.

All of the plant compounds that affected these nerves were produced by plants to protect themselves from being eaten (a cow that eats a chili pepper will think twice about eating these brightly colored plants again!). These compounds are all chemically similar to vanillin, the plant compound from vanilla beans that gives the flavor to ice cream. So, the proteins that bound them were initially called vanilloid receptors. It is curious that the study of these food flavorings has led to the discovery of how sensory neurons function.

Supraoptic neurons possess a specialized form of the TRPV1 protein that is also found in heat-sensing cells. This form, however, is chopped off on one end, making it shorter and insensitive to heat. Instead of reacting to heat, this protein responds to different degrees of stretch imposed upon the cell membrane. When a supraoptic neuron shrinks from dehydration, its membrane becomes wrinkled and activates the TRPV1 protein. This is how supraoptic neurons sense dehydration.[16]

Thirst and Alcoholic Drinks

Dehydration is not the only stimulus that affects how much ADH is made by supraoptic neurons. Most of us who have drunk alcohol-containing drinks have noticed that such drinks stimulate the production of urine in volumes all out of proportion to the volume of beverage consumed. This is because alcohol inhibits the secretion of ADH from the supraoptic nucleus. After an alcoholic drink, the kidneys produce larger volumes of a dilute urine and stimulate the need to drink more water to replace the water lost in urine.

Normally, supraoptic neurons rapidly recover from the effects of alcohol, and the water balance in the body is easily restored. However, there is some evidence that people suffering from chronic alcoholism may experience permanent damage to the supraoptic nucleus.[10] This evidence was obtained from a study of hypothalamic tissue from people who drank 100 g of alcohol per day (equivalent to 3.4 ounces of pure alcohol, or about four 2-ounce drinks of whisky). Hypothalamic sections taken from the brains of these people showed a significant loss in the number of neurons in the supraoptic nucleus, possibly indicating a permanent impairment of the ability to control water balance.

What accounts for the pleasurable qualities of drinking water or alcohol or consuming foods? A collection of neurons just anterior to the hypothalamus itself seems to be involved in the drive to consume food, water, alcohol, or even various types of drugs. This collection of neurons, called the nucleus accumbens, projects to the hypothalamus and controls the activity of feeding- and drinking-regulating nerve cells.[12,14] Addictive properties of tasty foods and drinks seem to be regulated by the nucleus accumbens, which permits the hypothalamus to initiate consumption when a food item is sufficiently rewarding. If the activity of specific hypothalamic neurons (containing agouti-related peptide, see chapter 1) is blocked, mice can be prevented from undergoing binge drinking of alcohol solutions.[17] Thus, the interrelationship between the hypothalamus and the nucleus accumbens is being eagerly studied in the hope that some types of addictive, consumptive behaviors can be corrected in human beings as well as in mice.

The Detection of Salt

All of us have experienced a craving for salty foods like pretzels or potato chips at some times but an aversion to them at other times. What accounts for this?

A decrease in blood volume and a lowered water content of the blood are not the only stimuli detected by the hypothalamus that can affect appetite and eating behavior. An increase in the salt content of the blood alone can cause an aversion to salt. This can be easily demonstrated.

Normally, mildly salty solutions are preferred by rodents and are eagerly lapped up in comparison to distilled water. For example, a rat seems to enjoy drinking a 0.1 molar solution of sodium chloride (this amounts to about a 0.6% solution, which is less salty than normal blood). A similar appetite for sodium can be seen in wild animals like deer, who will frequently visit a "salt lick" of salty rocks to replenish their body stores of sodium.

However, when a rat is denied water for 10 to 12 hours, becomes mildly dehydrated, and consequently has an elevated salt concentration in the blood, the rat will reverse its preferences and drink distilled water rather than a saline solution. Of course, it would make sense for a dehydrated rat to avoid salty solutions, but what causes this behavior? A logical conclusion is that some sort of a sodium sensor exists in the brain. However, where is it?

Experiments conducted over the last 10 years have identified yet another area of the brain with leaky capillaries that seems to contain the salt sensor. This area, called the subfornical organ (SFO), is located just above the thalamus and sends many neural projections to the hypothalamus.[4] Nerve cells in this area become highly active when blood levels of sodium rise above the normal level. It turns out, however, that the nerve cells of the SFO are *not* the sodium sensors. In fact, an entirely different kind of cell, the glial cell, carries out this function.

Glial cells extract nutrients from the blood, maintain the blood-brain barrier, and produce a layer of insulation called myelin that is wrapped around nerve axons.[23] Glial cells called astrocytes appear to be particularly important for sodium sensing. These cells possess long cell processes that contact blood vessels and nerve cells (these long, radiating processes give the cells a star-like shape that is responsible for their description as "astrocytes") (Fig. 2-2).

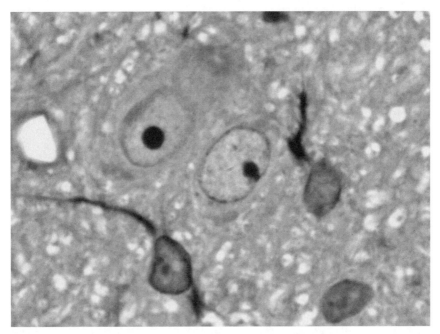

Figure 2.2. Astrocytes are accessory cells that possess pale, ovoid nuclei and long cell processes (stained here with an antibody to glial fibrillary acidic protein). They frequently contact neurons (cells with large, round nuclei and dark-staining nuclear structures called nucleoli) and blood vessels.

When sodium leaks out of the leaky capillaries of the subfornical organ, it passes into astrocytes via a specialized type of sodium channel located within the cell membrane. This uptake of sodium stimulates the astrocytes to transform glucose into a smaller molecule called lactate. Lactate, exported from the astrocytes, then stimulates adjacent neurons to fire, which causes an aversion to sodium.[20] The importance of this type of communication between astrocytes and nerve cells has become increasingly appreciated in recent years and will be discussed again later on.

THE HYPOTHALAMUS AND THE CONTROL OF BODY TEMPERATURE

The control of body temperature can often be closely linked with the control of water balance: when we enter a hot desert, we immediately yearn both for a cool drink and for a way of cooling off and lowering our body temperature. It is not too surprising, then, that the control of body temperature also seems to be a major task for the same areas of the anterior hypothalamus that are also involved in the control of thirst.

Normally we do not notice this vital function of the hypothalamus very much, in part because it is so efficient and successful: our body temperatures do not vary much from 98.6° F (37° C) over the course of the day. The control of body temperature can only be studied well when it varies from the norm. A dramatic resetting of our hypothalamic "thermostat" occurs in two circumstances: during a fever (in humans) and during hibernation (in animals like chipmunks and bears).

Fever

We all are familiar with the increase in body temperature (fever) that occurs during an infection of some sort. Fever is caused by an altered function of the hypothalamus, so when a doctor inserts a thermometer into your mouth, he is indirectly probing the activity of temperature-sensitive neurons in your hypothalamus.

It is generally believed that an elevated body temperature helps the body fight off an infection. However, fever is not a direct effect of the infection but stems from molecules produced by cells of the immune system that are activated to fight the infection. One of these molecules, called interleukin-1, travels up to the leaky capillaries of the OVLT of the anterior hypothalamus, enters the brain, and activates neurons of the preoptic area. It does not do this directly; instead, it causes cells in this area to produce an enzyme (cyclooxygenase) that synthesizes a lipid molecule called prostaglandin E.

Prostaglandin E is the culprit that activates hypothalamic neurons and induces fever. It is a remarkably potent molecule: infusion of as little as one nanogram (one millionth of a gram) of prostaglandin into the hypothalamus causes the appearance of fever. One of the reasons why we take aspirin to combat a fever is that aspirin is an inhibitor of cyclooxygenase. In the presence of aspirin, hypothalamic cells cannot make prostaglandin E, and thus the ability to produce a fever is diminished.

How do these few hypothalamic neurons manage to raise the temperature of the entire body? There are two ways. One mechanism involves projections of hypothalamic neurons to the medulla of the brain. Nerve cells in the medulla innervate so-called motor neurons of the spinal cord that control muscle contractions. When the hypothalamus is stimulated by prostaglandin E, it turns on a circuit that forces the spinal motor neurons to fire rhythmically. This causes most of the muscles of your body to rapidly contract and relax ("shivering"). These rapid muscular contractions are not only tiring but also generate a lot of waste heat that heats up the body, as those of us who walk to work or school on a cold winter morning have often experienced.

Another mechanism, directed through the paraventricular nucleus of the hypothalamus, activates nerve fibers of the autonomic nervous system that innervate a special population of fat cells. There are actually two different types of fat cells in the body, and only one type can generate heat.

Most of the fat cells in the body are enormous spherical cells that contain a large droplet of fat (lipid). These types of cells are called white fat cells and as they fill up with fat can become the giants of the world of cells. White fat cells can achieve a diameter of about one-tenth of a millimeter or so (100 micrometers). This may not seem like much, but since most cells are about one-tenth this size, fat cells really are impressive. They are almost large enough to be seen with the naked eye, aided by a strong magnifying glass. I have looked at the subcutaneous fat of an unconscious rat through a dissecting microscope, and this tissue provided an enchanting look at the world of cells. Under low magnification, I could see transparent fat cells tethered together into bundles like bunches of glistening balloons; each cell was filled with a droplet that looked like a bubble of salad oil. Capillaries containing thin streams of red blood cells rushing back and forth penetrated the spaces between the fat cells to provide them nourishment and to transfer lipids from the blood. I'll never again think about fat cells without remembering their beautiful appearance under the microscope.

The main job of white fat cells is simply to store nutrients for possible later use during a food shortage, and also to provide a layer of insulation beneath the skin. However, a second, sparser population of fat cells called brown fat cells also exists in the body, and these are the ones that are activated by the hypothalamus to generate heat.

Brown fat cells contain traces of some type of brownish pigment in their cytoplasm that makes this type of fat faintly brown colored. More importantly, these fat cells contain many small droplets of fat instead of a big one, and when the cells are activated, they withdraw fat from these tiny droplets and rapidly oxidize, or "burn," it (Fig. 2-3).

Fat in these cells is oxidized in cellular organelles called mitochondria. Most cells in the body contain at least a hundred of these tiny, sausage-shaped organelles; brown fat cells contain several times more mitochondria than that. Each mitochondrion is unusual because it is composed of an outer membrane, an intermembrane space, a second membrane, and a final interior space called the matrix. All other cytoplasmic organelles possess only a single membrane. The possession of two membranes by mitochondria is critical for their ability to generate energy and heat.

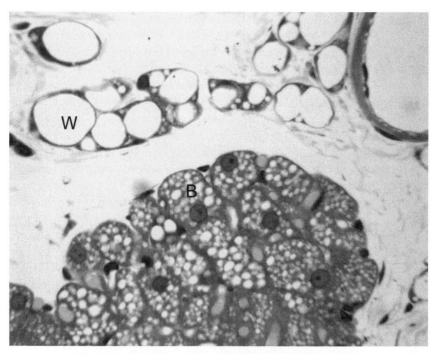

Figure 2.3.   White fat cells (W) have a huge, single drop of lipid in their cytoplasms and a flattened cell nucleus that is displaced to one side by all the fat. Brown fat cells (B) have a single round nucleus that is surrounded by smaller droplets of fat.

To burn (oxidize) nutrient molecules such as fats (lipids), lipid molecules are first transported from cytoplasmic fat droplets into the mitochondria via a pore in the outer mitochondrial membrane formed by a protein called the voltage-dependent anion channel. Then enzymes within the mitochondrion break up the large lipid molecules into smaller molecules and combine them with oxygen. As they do this, protons (hydrogen ions) and electrons are stripped off of the nutrient molecules. The hydrogen ions are shuttled into the intermembrane space between the outer and inner mitochondrial membranes. They acquire high concentrations in this space and create a strong charge separation between the intermembrane space and the interior of the mitochondrion.

Naturally, Mother Nature wants to disperse this charge separation, just as separated ions within a battery want to travel down a circuit to meet each other and do work. This is done in mitochondria by forcing the hydrogen ions to travel into the matrix via a protein pore called the ATP synthase protein, which is embedded in the inner mitochondrial membrane.

The ATP synthase protein is a remarkable little machine. It is composed of three main parts. One part allows hydrogen ions to flow into an opening at the bottom of the protein. As the hydrogen ions enter, they are transferred to a wheel-shaped subunit (second part of the protein) that rapidly rotates like a merry-go-round within the plane of the cell membrane. As more and more hydrogen ions attach to the wheel, they cause the wheel to spin faster and interact with a third part of the protein. This third part responds to being buffeted by the wheel by converting a molecule called ADP (adenosine diphosphate) into the energy-rich molecule ATP (adenosine triphosphate). When carried throughout the cell, ATP molecules add high-energy phosphate groups to many proteins (enzymes, motor proteins, etc.) and provide the energy needed to operate the cell. This is a relatively efficient process that operates in all cells.

This process is different in brown fat because the mitochondria in these specialized fat cells have been engineered to be very inefficient in oxidizing fuels. This is because they have a specialized protein, called an uncoupling protein, that wastes the energy obtained from burning fat. The uncoupling protein opens a pore between the intermembrane space and the mitochondrial matrix and permits hydrogen ions to flow back into the matrix without having to pass through the ATP synthase protein and without generating any ATP. This flood of hydrogen ions pouring through the membrane and into the matrix generates heat, just as a current of electricity flowing through a small wire encounters resistance and causes the wire to heat up. This is the mechanism whereby brown fat can generate waste heat to warm the body.

Brown fat cells are generally located along the major blood vessels that supply the brain with blood, and they play a role in warming the blood going into the skull so our brains don't freeze on a cold winter morning. Brown fat cells can also warm up other parts of the body in a process called "nonshivering" thermogenesis. In addition, brown fat cells are activated by the hypothalamus during a fever.[3,15] It has been calculated that brown fat accounts for as much as 10% to 20% of all the utilization of energy in the body.[22]

Both types of fat cells can thus help keep us warm in the winter, either by adding a layer of insulation beneath our skin (white fat cells) or by undergoing nonshivering thermogenesis (brown fat). So, although fat cells have begun to be regarded as our enemies—because of the growing problem of obesity—they really have a very important role to play in the body.

Nothing demonstrates the importance of fat cells better than experiments with mice that were genetically engineered to lack the ability to produce fat cells. This can be done by inactivating the gene for a protein called C/EBP that binds to DNA and which normally turns on other genes needed for the maturation of fat cells from precursor cells called preadipocytes. Molecular biologists commonly can achieve this by genetically manipulating an egg cell retrieved from a donor mouse and then implanting the fertilized egg into the uterus of a recipient mouse. When a litter of mice is born from the recipient mouse, the baby mice have no fat cells at all in their bodies! This may seem like a dream come true for us dieting humans, but in fact it has disastrous consequences for the mice.

When mice lacking fat cells are young and still drinking mother's milk, they always have a rapidly digestible source of energy right within their stomachs and so have a normal metabolism. However, when they are weaned and eat more slowly digestible solid food, they have no way to store energy-rich molecules to power their muscles and body because they have no fat cells. Thus, when fasted, they have no quick access to sources of emergency calories. Such mice rapidly use up the glycogen stored in the liver after a meal; as a consequence, the liver cannot convert glycogen into glucose, blood glucose drops to very low levels, and the mice "run out of gas." Muscles cannot use either blood glucose or lipids released from fat cells for energy and thus cease functioning. The brain, which usually depends upon the breakdown of glycogen into glucose for energy, undergoes a drastic drop in activity, and the mice drop into a state not unlike a coma. This condition is termed severe torpor. The mice stop running around and have body temperatures that plummet to close to air temperature until they manage to finally digest the solid food in their stomachs and wake up again.[6] These mice show us why we must be thankful for the fat cells that we have, in spite of the other problems that they may cause.

The hypothalamus provokes elevations in body temperature (fever) in response to other blood-borne molecules in addition to interleukin-1. Another protein, called RANK, has also been recently found to have a commanding role in the generation of fever.

RANK is a protein found in the cell membrane of many different types of cells. RANK acts as a receptor for a circulating protein called RANK-ligand. Originally, RANK was discovered due to its potent effects upon the resorption and destruction of bones by specialized cells called osteoclasts. Thus, it was very surprising to find that RANK and RANK-ligand also exist within the brain as well as in the bones.

When RANK-ligand is infused into the hypothalamus and activates its RANK receptor, mice develop fevers of 103° F! However, when mice are genetically altered to lack RANK, they do not develop fevers. Curiously, restoring RANK to astrocytes is sufficient to restore the fever response to RANK-ligand, so astrocytes once again must be responsible for the activation of fever by RANK.[9] Apparently, astrocytes respond to RANK by producing prostaglandins, which stimulate nearby neurons to induce fever.

An "experiment of nature" found in humans has validated the experiments about RANK and fever that have been performed on mice. A rare inherited mutation in the RANK protein has been discovered in a small family in Turkey. These individuals have an inactive RANK protein and display abnormally thickened bones and fail to develop fevers after contracting pneumonia.[9] Thus, information obtained from mice seems once again to instruct us about hypothalamic function in humans.

Hibernation

Hibernation is defined as an event in which activity, metabolic rate, oxygen consumption, and body temperature all plummet during the winter. In animals such as chipmunks, body temperature can fall from about 85° F to as low as 46° F (29° C and 8° C, respectively) when they are "asleep" within their underground burrows. During hibernation, animals are somehow protected from a variety of harmful influences: hypothermia, which can destroy cells; muscle disuse, which normally results in muscle atrophy and degeneration; and even tumor growth, which can be fatally damaging to animals that are not undergoing hibernation.[11] Hibernating animals can reduce their metabolic activity to as low as 2% of normal.[3] How these marvelous effects of hibernation protect the cells and tissues of an animal, and how an animal's entire metabolism is adjusted during hibernation, are questions that have still not been answered satisfactorily.

Chipmunks initiate hibernation during the winter when the length of the day becomes shorter than the length of the night. As we shall see in chapter 4, a hypothalamic nucleus called the suprachiasmatic nucleus processes information about day length and must be responsible for initiating hibernation.

Other animals, like hamsters and mice, may also undergo hibernation, but these animals seem to do this primarily in response to the decrease in food availability during the winter months. Some strains of mice react to diminished food and bodily fat stores by going to sleep and reducing their body temperatures to about 55° F. However, if these mice are infused with leptin, they stay awake and warm even in the absence of food. Thus, the ability of leptin to function as a signal of calories affects both the hypothalamic control of feeding and the hypothalamic control of body temperature and sleep. The specific site in the hypothalamus affected by leptin has been identified by a simple experiment: if hamsters are given monosodium glutamate, which damages the arcuate nucleus of the hypothalamus but leaves other areas intact, they will no longer hibernate in response to decreases in food and fat stores.[6,18]

Curiously, a newly discovered protein called hibernation specific protein has been found to be critical for the initiation of hibernation in chipmunks. This protein is produced by the liver and secreted into the bloodstream to reach the hypothalamus. If antibodies to this protein are injected into a chipmunk to block its effects upon the brain, the chipmunk will fail to go into hibernation.[11] Further study of this protein may identify the root mechanisms of hibernation and could perhaps have medical uses as well.

The Evolution of Warm-Bloodedness

We take the maintenance of a stable and high body temperature for granted, but in fact this ability, termed endothermy, was a major event in the evolution of vertebrate life on earth. Many of our vertebrate ancestors, including turtles, frogs, and most reptiles, had a "cold-blooded" (ectothermic) metabolism that is unlike our "warm-blooded" (endothermic) physiology. When a reptile like a crocodile wants to become more active, he is forced to raise his body temperature by basking in the sun until he heats up. Mammals and birds do not generally need this, because they are continually generating extra heat and because the hypothalamus constantly regulates this heat production to maintain a stable body temperature.

The advantage of endothermy is that mammals are generally prepared for anything and can dash off in search of a meal or a mate at any time, day or night, even if the ambient temperature is low. This allows mammals to occupy many environmental niches that are denied to cold-blooded creatures. The advantages of endothermy, however, do not come without a price.

The disadvantage of endothermy is that mammals must maintain a metabolic rate that is 5 to 10 times higher than that of reptiles.[22] To do this, mammals must find and efficiently digest a lot more food each day than an average reptile. Almost every aspect of mammalian anatomy is devoted to this task of finding food and extracting enough calories from it for a warm-blooded mode of living.

Mammalian teeth, for example, are far more specialized than those found in reptiles. Crocodiles have rows of identical cone-shaped teeth that are great for holding a prey animal before it is unceremoniously dumped down the gullet for a prolonged period of digestion. Mammals, in contrast, have a variety of tooth shapes: incisors for nipping and cutting, canines for biting, and molars for grinding. These types of teeth allow mammals to crush food much more effectively than do crocodiles and allow us to more efficiently extract each and every calorie from every bite of food.

Our noses are also specialized for endothermy. Mammals have scroll-shaped bones called nasal turbinates that hang from each side of the nasal cavity. These bones are covered by a highly vascular mucosal layer that has lots of warm fluid in it. When cold air from outside passes along the convoluted pathways between these bones, the air becomes warmed and humidified before it enters the trachea and the lungs. This prevents the cold air from drying out our warm lungs. One of the current controversies in biology is whether or not turbinate bones can be found in dinosaur skeletons and if this might mean that some types of dinosaurs could have been warm blooded.

Another obvious modification seen in mammals is that most of them are covered with an insulating layer of fur to diminish heat loss from the body during cold weather (larger animals like elephants need less hair because of their thick layer of insulating skin). On a cold day, the hypothalamus activates the sympathetic nervous system, which directs tiny muscles attached to hair follicles to pull hairs erect (even humans experience this as "goose bumps" evoked by chilly temperatures). By fluffing out the hairs of the skin, the hypothalamus traps lots of air within the furry coat of squirrels and rabbits and other animals and thus improves the insulating qualities of the skin.

During warm weather, the hypothalamus also affects the ability of the skin to regulate body temperature: on a hot day, the hypothalamus directs the autonomic nervous system to stimulate eccrine sweat glands beneath the skin that provide a means of cooling off by evaporation of sweat from the skin. Reptiles, of course, lack all of these thermoregulatory organs.

A final distinction between reptiles and mammals is that mammals possess mammary glands, the organs from which the name "mammal" is derived.

These organs allow mammals to nourish their metabolically active and very hungry young. Even this feature directly springs from the need to maintain a stable body temperature: mammary glands actually represent highly modified sweat glands that evolved into milk-producing organs.

Mammals are not the only warm-blooded creatures on earth. Birds are also highly active metabolically and maintain stable elevated body temperatures. They use feathers instead of hairs for insulation; they also completely lack brown fat and must generate their body heat by the inefficient burning of fuels by muscle cells and also by shivering.[2,22]

How much does the evolution of warm-bloodedness owe to changes in hypothalamic function? While the basic structures that make warm-bloodedness possible—hair, a high metabolism, specialized noses, and so forth—require drastic changes in the overall anatomy of the body, they still would be poorly effective if they were not regulated by the hypothalamus. It would do no good to fluff out your hair on a hot day or sweat during a snow-storm! Did the evolution of a temperature-sensitive hypothalamus precede the development of other mammalian features? This does seem to be true. Animals like frogs or snakes can't generate much of their own heat, but they can move to warmer environments or bask in the sun if they get too cold. The hypothalamus of reptiles and amphibians is broadly similar in structure to that of mammals, and if the preoptic area of the hypothalamus is damaged, these animals fail to undergo behaviors that protect them from cold.[1] So, a temperature-sensitive hypothalamus appears to be a very ancient part of the brain. During the evolution of mammals, it acquired new connections with the autonomic nervous system that allowed it to direct muscles, the skin, and brown fat to maintain a stable body temperature.

Changing the Hypothalamic Thermostat

Temperature regulation by the hypothalamus can be studied by traditional means used by neuroscientists, for example, by squirting chemicals into the hypothalamus that affect neuronal firing rates or by damaging temperature-sensitive neurons. These techniques are useful, but they do not really mimic what is going on in an undamaged hypothalamus. More subtle, newer techniques have now arisen due to the ingenuity of molecular biologists. One such technique recently revealed startling information about the hypothalamus and thermoregulation.

The technique I am referring to is genetic engineering, which can create animals with modified chromosomes and modified genes. To perform genetic engineering, an egg cell from a mouse is injected with artificially created molecules of DNA. One of these DNA molecules can be very similar to the normal stretch of DNA on a chromosome that codes for a particular protein. Most of the time, the artificial DNA introduced into the egg cell will just be destroyed; but occasionally it will pair up with the gene it was meant to replace, and the egg cell will switch it with the normal gene and incorporate it into the chromosome. As a consequence, this egg cell will carry an artificially modified gene; when fertilized by a sperm and implanted into the uterus of another mouse, it will develop into a baby mouse with a modified gene. This mouse can be bred with other mice to produce a whole colony of genetically modified mice. This is the approach that was used to change the temperature sensitivity of the hypothalamus in mice.

The specific experiment that studied hypothalamic thermoregulation took advantage of the fact that in the middle of the hypothalamus, a cluster of about 4000 neurons exists that all utilize a protein called orexin as a neurotransmitter (see chapter 4 for more discussion of these cells). No other neurons in the brain make this protein. One of the reasons for this is related to the structure of the gene for orexin. It is composed of a region of DNA that codes for the protein (the gene itself) plus a preceding DNA region called a promoter. A number of DNA-binding proteins adhere to this promoter region and "turn on" the associated gene for orexin. One such DNA-binding protein, called nuclear receptor 6A1, seems to be made mainly in orexin-containing neurons and is responsible for the fact that only these cells make orexin.

Bruno Conti and coworkers at the Scripps Institute in San Diego were interested in modifying the function of these orexin-containing neurons.[2] To do this, he created a genetically modified mouse. The artificial gene he inserted into the mouse contained the promoter for orexin, but it was fused to the gene for brown fat uncoupling protein instead of the gene for orexin. Thus, when the modified mice grew up, their orexin-containing neurons in the hypothalamus made the brown fat uncoupling protein along with the orexin. What was the result of this?

When the orexin-containing neurons in the hypothalamus made the uncoupling protein, this forced their mitochondria to burn nutrients inefficiently and thus generate heat, just as if they were brown fat cells within the hypothalamus. In effect, each cell acted like a tiny candle. These hot little candle cells were located only 0.8 mm posterior to heat-sensitive neurons of the anterior hypothalamus. The thermosensitive cells near the hot orexin-containing cells reacted as if the blood entering the hypothalamus was too hot. This caused them to cool down the entire body of each mouse by about 0.5 degrees C (or about 1 degree F). The whole experiment was kind of like holding a candle up to a thermostat mounted in the hallway of a house: by warming up the thermostat, a person could cause the thermostat to shut down the furnace earlier than normal, resulting in a lower overall temperature in the house.

All of these cooled-down mice appeared to act normally in almost all respects, though they tended to get a little chubby because, even though they ate a normal amount of food, their lowered metabolic rates and lowered body temperatures allowed them to use food more efficiently. More importantly, their lowered body temperatures were associated with a 20% increase in lifespan! The slower metabolic processes that occurred in the cooler mice apparently generated less damage to cell constituents than normal, resulting in a slowing of aging. The only other thing known to genuinely slow aging is caloric restriction: from the research on rats and monkeys conducted so far, cutting daily calorie consumption by about 30% from normal also seems to prolong life. It's almost like the cooled-down mice came to resemble those giant tortoises found in the Galapagos. The slow-moving and cold-blooded tortoises live much longer than an average mouse because their metabolism is so low. So, the imposition of a high metabolic rate and a high intake of calories by the hypothalamus has a significant effect on longevity, putting the "fountain of youth" just that much farther out of reach for us.

Body Temperature and Fertility

Complicated experiments with modified genes are not the only way of resetting the "thermostat" in the hypothalamus. Mother Nature has provided us with another, much simpler way. It turns out that the thermoregulatory neurons of the hypothalamus possess receptors for steroid hormones that alter their activities. The female sex hormone, estrogen, lowers basal body temperature slightly by acting on these cells, whereas another hormone, progesterone, elevates basal body temperature.

Blood levels of progesterone spike at a particular time in women each month: during the process of ovulation, when the release of an egg cell from the ovary transforms the cells that had surrounded the egg into a new structure called a corpus luteum. The new job for each newly formed corpus luteum is to secrete progesterone. Thus, if a woman wants to have a good idea when ovulation takes place, one way is to carefully measure body temperature each day just after waking. A sudden elevation in body temperature by about 0.5 degrees C is a fairly reliable way to pinpoint the day of ovulation. This information can be used with some success to avoid pregnancy, as long as sexual intercourse is avoided prior to ovulation. This is the basis for the "rhythm method" of contraception that was popularized by a Catholic priest in Germany, Wilhelm Hillebrand, during the 1930s. This method, unlike other chemical or barrier methods of contraception, has been adopted as morally correct by the Catholic Church. Its only drawback is that it requires a very disciplined approach to measuring temperature and modulating behavior that may be difficult for some individuals to maintain.

The Hypothalamus and Other Responses to Infections

The generation of fever by the hypothalamus during an infection is not the only response provoked by illness. You and I can readily testify what these responses are: when I get the flu or another infection, I not only develop a fever, but I also feel achy, lethargic, and sleepy, and I lose my appetite. In fact, if I didn't feel so miserable, getting sick would be a great way for me to lose weight! Why do these other responses to illness occur?

The lack of appetite that we experience during illness seems to be due to the production of so-called cytokine molecules by cells of the immune system. Interleukin-1, the cytokine that induces fever, can also bind to appetite-restraining cells in the arcuate nucleus of the hypothalamus that utilize proopiomelanocortin (POMC) as a neurotransmitter. In mice with functional receptors for interleukin-1 on these cells, illness induces a fall in food intake, but if the receptor is deactivated, ill mice eat normally.[7] Also, mice and men both tend to sleep much more during an infection. This makes sense, because the bodily energy that otherwise would be expended in walking or playing soccer now can be used to fight a disease. Once again, activation of the hypothalamus by cytokines seems responsible, and in fact it involves sleep-regulating neurons in the lateral hypothalamus that use orexin as a neurotransmitter.[8]

The Hypothalamus and the Control of Sex and Emotions

In conditions of extreme heat or famine, the hypothalamus focuses our attention simply upon staying alive by drinking water, cooling off, or finding food. However, in more normal circumstances, we can indulge in less urgent drives. One of these is the drive for reproduction and sexual behavior.

Sex is all around us: as the song says, birds do it, bees do it, and more generally, almost all complex organisms on earth, from fish to flowers to trees to octopuses, engage in the sexual mode of reproduction. Sex ensures that the offspring of two individuals are genetically different from either parent. This results in a tremendous diversity of individuals within a species, even if they all look rather similar to each other. It is this valuable genetic diversity that makes sex so important in the broader scale of things.

To most of us, however, sex is important not because it improves our species but because it has a tremendous impact upon our lives. The quest for a mate and the physical and emotional pleasures of sexual relationships preoccupy us for many years of our lives and form the underlying themes in most of human literature. Thousands of novels have been written about the quest for love in humans, but none have been written about the quest for the perfect steak or the most appetizing drink (even though cookbooks do exert a claim upon our time). The hypothalamus has a major role to play in our searches for love and sex.

Most of us tend to be rather shy about our sex lives. We tend to talk volubly about where we spend our vacations or which restaurants we prefer, but few of us discuss what we did between the sheets the previous night. Why are we so shy about sex? Partly, I think, because we are all told as little children that we must keep our clothes on and be private about our bodies.

Since sex most often must be performed with our clothes off, this might be one reason we are so reticent about it.

This reticence gets in the way of a full understanding of the characteristics of human sexual behavior. Openly asking each other about our sex lives doesn't happen very often. Instead, one way of learning about human sexual behavior is via the publication of surveys of thousands of people that pose questions about sex that we rarely ask ourselves.

Some of the first large-scale surveys about sex in the United States were published by Alfred C. Kinsey in 1948 and 1953. This detailed analysis of 17,000 respondents revealed information that was shocking for the times; for example, about 80% of males admitted to having had sex before marriage, and 50% of married men admitted to having extramarital affairs. A more recent update of this approach was published in 1993.[21] The data in this survey reveal that humans are ingenious about sex and apply a variety of approaches to seek sexual satisfaction.

While most of us are engaged in conventional heterosexual relationships, a surprisingly high percentage of adults reported that they had experienced at least one homosexual encounter (22% of males and 17% of females). However, smaller percentages of the population stated that they were engaged in multiple or ongoing homosexual relationships (9% of males and 5% of females). About 90% of men and women feel that oral stimulation of the genitalia is a normal part of a sexual relationship. Most of us (89% of men and 75% of women) find that the best way to achieve sexual satisfaction is through intercourse, while smaller numbers of people prefer manual stimulation by either their partner or by themselves.

Significant numbers of people engage in less conventional sexual behavior, such as threesomes (14% of married men). About 20% of married men admitted to having had commercial sex with prostitutes. There is a disturbing connection between sexual abuse and prostitution: almost 80% of female prostitutes reported that they had been sexually abused before the age of 12; as a consequence, many prostitutes have an impaired ability to enjoy sex and only engage in it for monetary reasons.

Another disturbing finding is that 23% of all women reported experiencing some type of sexual molestation as children. About 2% of men reported having had sex with children. I rather believe that it's no business of mine what anyone's sex life is like, as long as no harm is being done to anyone and as long as it's voluntary. These types of behavior, however, violate these basic principles.

Certain stimuli that have no erotic effects on most of us can be sexually rewarding to others. About 10% to 15% of people reported that mental humiliation or physically painful stimuli were sexually stimulating for them.

Smaller numbers of people have sexual fetishes; for example, they find women's feet to be more sexually stimulating than other portions of the body, or they react positively to certain silk cloth or other textures.

What explains the sexual attraction of one person for another, and what accounts for the variability in sexual behavior noted above? Why are men primarily attracted to women and vice versa? Which features make a woman more sexually attractive to a man, and which features of a man attract women?

It is true that beauty is in the eye of the beholder and that many features of a person can be attractive. However, common sense, coupled with a number of scientific surveys on the subject, tell us which aspects of a person seem more important in sexual attractiveness. Men generally report that women with a small waist-to-hip ratio, longer legs, and larger breasts are most sexually attractive, while women seem to prefer men with a greater height, larger overall size, and shorter legs relative to their height.[5]

Most of us strive to be as attractive as possible, but there is not a whole lot any of us can do to optimize most of these bodily features. High heels may make women's legs look longer, and shoe inserts may give men the illusion of height, but genes seem to play the most important roles in how these physical attributes develop and there is no way of really altering them short of major surgery. Even breast size seems genetically determined: in a recent study of fraternal and identical twin women, it was estimated that genes account for almost two-thirds in the variability of bra cup size.[39] About the only thing most of us can do to change our bodies, unless we undergo plastic surgery, is to keep our levels of body fat within a range that allows for an appealing body shape. This is probably the main motivation for dieting for most of us, in addition to more mundane and abstract reasons for improving one's health.

I suppose the continual quest for physical perfection is a natural and very human thing, but personally, I feel it is overrated. To me, a welcoming smile is one of the most attractive features a person can have. If we all needed to have perfect bodies and beautiful faces to experience a satisfying sex life, the human race would be extinct by now! When I watch TV shows or movies from the 1940s and 1950s, I find it to be a relief to see relatively normal-looking faces with a pleasing variability rather than the ruler-straight noses, wide-open eyes, and puffy lips produced by the surgeon's knife that are so often on display now. However, my personal prejudices seem to represent a minority opinion.

How does the brain translate signals of sexual attractiveness into cues for sexual activity? Many parts of the nervous system contribute to the expression of sexual behavior, but it is clear that the hypothalamus has a particularly prominent role to play in sex.

MALE SEXUAL BEHAVIOR

The contributions of the hypothalamus to sexual behavior have been studied extensively in rats and other rodents that act as models for studying sexual behavior in humans. In order to undergo male sexual behavior, a rat must (1) locate a sexual partner, (2) elicit sexual receptivity in that partner, (3) experience sexual arousal and penile erection, and (4) undergo copulatory (mounting) behavior.

To study sex between rats, pretty rudimentary equipment is required. First, you need male and female rats. Preferably, the male rats can be of a slightly different strain than the females to help tell the two apart more easily. For example, in our lab, we frequently utilized male "hooded" rats with black areas on their white coats of fur. When they approached female albino rats, which are completely white in color, it was easy to tell the males from the females, even at a distance. This is important because rats are nocturnal animals and won't behave for an observer unless the ambient lighting is subdued (low-wattage red lightbulbs are best). So, usually we placed the two rats into a large, transparent Plexiglas box containing sawdust to catch their droppings. If they became interested in each other, we carefully peered at them in the dim red light and, using a stopwatch, timed how long it took for them to get together. Occasionally, as graduate students, we would stop our work, look at each other, and wonder aloud how we ever got into such a crazy business as watching rats have sex. The things people do for science! Studying sexual behavior in rats sounds lascivious, but actually it could become quite tedious if large numbers of animals had to be analyzed.

The whole procedure for studying sex behavior in rats is standardized and is commonly used by hundreds of researchers interested in the hypothalamic control of sex. A number of behavioral attributes are routinely studied. The amount of time that passes before a male rat climbs onto the back of a female is called the mount latency. The number of times a male displays a pelvic thrusting motion is called the intromission frequency. Other measurements (ejaculatory frequency and latency, etc.) also fall into the range of behaviors examined.[45]

All of these phases of behavior are strongly influenced by male hormones like testosterone, and some aspects of these behaviors can be observed in both male and female rats under the right hormonal conditions. In fact, to be sure that the female rats would be receptive, we routinely injected them with estrogen and progesterone, steroid hormones that have powerful stimulatory effects upon sexual behavior. Other combinations of sex steroids can have unexpected effects; for example, a normal female rat will mount a male rat and perform thrusting movements as if she were having intercourse with the male if treated with testosterone. All of these behaviors are much more stereotyped in rats than in humans and can be monitored and quantitated easily.[2]

Since the 1950s, it has been known that large lesions of the anterior hypothalamus/preoptic area virtually eliminate male sexual behavior in rats. On the other hand, a pioneering investigator named Paul McLean discovered that stimulation of the anterior hypothalamus with an implanted electrode caused penile erection in rats. Finally, it has long been known that neurons sensitive to testosterone are widespread throughout the hypothalamus. So the idea that the hypothalamus has an important role in the control of sex is not a new one. What hasn't been known with precision is which specific nerve cells are crucial for the control of male sexual behavior.

I was a witness to one interesting discovery in this field when I was a young graduate student in the lab of my advisor, Dr. Roger Gorski, at UCLA. Long ago (in the 1970s), a distinguished professor was invited to give a talk to our lab group. This professor, Dr. Fernando Nottebohm, studied singing behavior in birds. He noted that males of many species sing to court females and warn away male competitors, whereas female birds are either virtually silent or sing only a little. When several brain nuclei needed for singing behavior in birds were examined, it was found that males had substantially bigger nuclei than females and that the size of these clusters of neurons could be affected by injections of testosterone. He hypothesized that the sex difference in brain structures might explain the sex difference in singing behavior in birds. Now it has been recognized that this hypothesis is probably too simple and does not apply to all species of birds.[14] However, at the time, it was a remarkable proposal.

A postdoctoral student in our lab, Larry Christensen, wondered if a similar phenomenon could be found in the rat brain. He pulled boxes of glass slides off the shelf and laboriously looked through hundreds of brain sections to see if any nuclei looked different between males and females. Larry called us all in to look at what he found: he put a glass slide on a microprojector that showed an enlarged image of the brain sections on a piece of paper. Even using a magnification of only ×5 or ×10, it was clear that a small, unnamed cluster of cells in the anterior hypothalamus was much bigger in males than in females (Fig. 3-1)![9,16]

How could the very anatomy of the hypothalamus be altered in males as compared to females? Could this sex difference in brain anatomy provide the explanation for sex differences in behavior in mammals?

This initial discovery by Larry led Dr. Gorski and many others to examine this sexually dimorphic nucleus (SDN) in rats and in many other mammalian species. Curiously, after making this discovery, Larry decided to leave the world of academic science and take up the study of Buddhism and other philosophies.

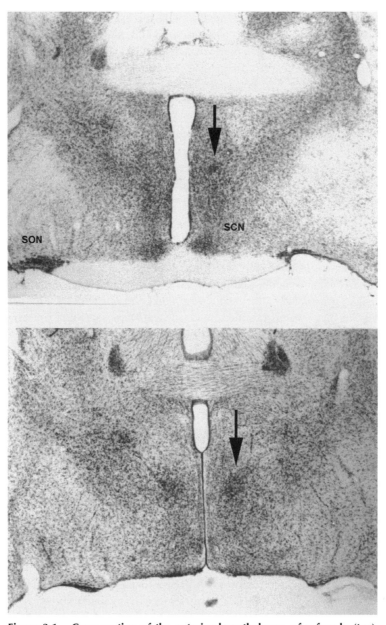

Figure 3.1. Cross section of the anterior hypothalamus of a female (top) and a male (bottom) rat. The small clusters of neurons on both sides of the hypothalamus (arrows) comprise the sexually dimorphic nucleus (SDN) of the hypothalamus, which is clearly larger in male than in female brains (reprinted from reference 43, with permission). The supraoptic (SO) and suprachiasmatic (SCN) nuclei are also visible in these brain sections.

He could have used his work as the foundation of an academic career, but he chose not to. I have often wondered about the path he chose. Becoming a neuroscientist does have its rewards: you have the opportunity to find out new information that no one else has found, you have the pleasures of viewing and examining beautiful and complicated regions of the brain, and there is always the hope that your work will in some way lead to the relief of human suffering due to neurological disorders. On the minus side, neuroscience requires a continual competition for scarce research funding, there is always a degree of acrimony in debates about scientific findings, and even if you can publish your findings in a journal, they are not always recognized by others. These pluses and minuses come with the job of academic science that Larry opted out of.

Study of the sexually dimorphic nucleus by many labs has led to answers to many questions about sexual behavior.

First, if baby rats are injected with testosterone during the first few days after birth, these injections permanently masculinize the size of the SDN in females and make females more prone to exhibit male-type sexual behavior. This seems to be due to sex differences in the deaths of neurons in the hypothalamus during development. Testosterone prevents many of the SDN neurons from dying, and thus the nucleus becomes bigger in males than in females. Conversely, early castration of baby male rats and removal of testosterone from their bodies diminishes the size of the SDN and permanently diminishes the expression of male sexual behavior later in life, even if testosterone is given to replace the hormones that would have been produced by the testicles of castrated rats. These data seem to support the idea that hormone-stimulated changes in SDN anatomy and function can cause changes in sexual behavior.

How would nerve cells in the hypothalamus alter behavior? A leading researcher of this question is Pauline Yahr at UC Irvine. She has found that many neurons within the anterior hypothalamus (and also neurons in a related area called the bed nucleus of the stria terminalis, which is located just above the SDN) project to another cluster of neurons located further down in the brain (the midbrain). These neurons, in turn, project to areas in the pons that control movement by regulating the spinal cord, and also to an area just anterior to the hypothalamus that regulates pleasure and addictive behaviors (nucleus accumbens). If this pathway from the anterior hypothalamus is cut, spontaneous male sexual behavior in rodents is greatly diminished.[36] Gradually, the circuitry that stimulates male sexual behavior is becoming known. Some simple portions of this circuitry are found in the spinal cord and are relatively easy to understand.

What simple circuits do I have in mind? It has been recently established that there is an "ejaculation center" within the upper lumbar spinal cord of rats. Nerve cells in this center are sensitive to testosterone and project to several other sets of neurons. One set of neurons in the lower lumbar cord sends stimulatory messages to the muscles attached to the penis. This set of neurons is called the spinal nucleus of the bulbocavernosus. Male rats have about 200 neurons in this nucleus, whereas female rats have only about 60 neurons.[34]

Other neurons innervated by the ejaculatory nucleus belong to sympathetic and parasympathetic ganglia and innervate the blood vessels of the penis, causing them to fill with blood during sexual activity and to provoke penile erection. There are sexual dimorphisms in these neurons as well. For example, male rats have three times as many of these sympathetic neurons projecting to the genitalia as compared with females.[12]

How do these neurons cause penile erection? They release a very peculiar neurotransmitter from their nerve endings that causes the smooth muscle cells of blood vessels to relax. The neurotransmitter—called nitric oxide—is released from nerve endings in the form of a gas. This gas then diffuses over to smooth muscle cells that wrap around blood vessels and stimulates them to make a molecule called cyclic GMP (guanosine monophosphate). It is this cyclic GMP molecule that causes penile blood vessels to relax, dilate, and cause an erection during sexual intercourse.

Since the basic circuitry for erection and ejaculation is found within the spinal cord, it is possible for men with spinal cord damage to perform these sexual functions. Even if an upper spinal cord lesion prevents sensory information from traveling from the pelvis to the brain, and even if descending stimulatory pathways are cut, manual stimulation of the genitalia can still elicit ejaculation. In this way, partially paralyzed men can still serve as sperm donors for wives who want additional children from their injured husbands.

The spinal neurons involved in all of this receive commands from the ejaculatory center of the spinal cord. These command neurons utilize a small protein called gastrin-releasing peptide (GRP) as a neurotransmitter; when GRP is released from the nerve endings of the command neurons, the GRP binds to the GRP receptors of the secondary clusters of neurons to set in motion all the major events that occur during copulation. These neurons also all require testosterone to be fully active; if rats are castrated to remove their source of testosterone, the neurons become lazy, and rats show a lowered ability to perform sexually. However, if GRP is infused into the spinal cord of a castrated rat, his sexual ability is completely restored, even if he no longer has much testosterone.[31]

All of these data show that there exists a chain of sexually dimorphic, testosterone-responsive neurons extending from the hypothalamus and down into the spinal cord that all contribute to the physiology of male sexual behavior. What is the role of the hypothalamic SDN in all of this?

One curious finding about the hypothalamic SDN of the rat is that it receives information from a part of the brain called the accessory olfactory bulb, which detects odors. How would the detection of odors be related to sex?

Detection of odors may be a key activity that solves the puzzle of how the SDN modulates sexual behavior. The sense of smell is a complicated thing: mammals have two systems for detecting smells.

The main olfactory bulb reacts to most of the things we smell. It contains neurons that receive information from the upper part of the nasal cavity; when activated, these neurons send impulses back to the brain via a long nerve called the olfactory nerve (labeled "O" in Fig. 1-1). The functional capacity of these cells to detect odors varies greatly from species to species. In mice, a family of over 1000 individual proteins has been detected that constitute the olfactory receptor family. Humans possess a much more limited variety of olfactory receptor proteins (about 350) and are less able to distinguish between various odors than animals like mice that depend heavily upon olfactory cues from their environment.

An activated neuron in the nasal cavity sends a signal to the olfactory bulb via an axon that terminates in only 1 out of about 2000 spherical structures that are called olfactory glomeruli. Within each glomerulus, initial processing of olfactory information occurs, and then this information is sent on to the cortex of the brain. A single cortical neuron can receive information from as many as 50 olfactory glomeruli, so a great degree of convergence of information takes place.[30] Presumably, a complex smell (burning wood, steaming soup), containing many different odorant molecules, activates numerous glomeruli that send signals to many different cortical neurons, leading to a very complicated interpretation of what is in the air.

The nasal sensory epithelia and olfactory bulbs are not the only olfactory organs. In many mammals, another organ called the vomeronasal organ plays a major role in reacting to specific odorants called pheromones that are secreted into the environment by animals. These odorous components of sweat or urine have a powerful influence on behavior in many mammals. Sensory cells of the vomeronasal organ are located in small pits associated with the nasal septum. In mice, each cell expresses 1 protein out of about 300 proteins that bind pheromones.[28] The vomeronasal cells project to the accessory olfactory bulb and other structures quite different from those involved in detecting nonpheromonal odors.

A range of behaviors are affected by pheromones and the vomeronasal system. Male mice will frequently attack a strange male placed into their cages; if the strange male is castrated, however, it will not be attacked. The involvement of testosterone-stimulated pheromones in eliciting attacks can be demonstrated by simply "painting" the back of a castrated male with the urine of an intact male, which will restore attacks upon the urine-painted male.

Another striking effect of pheromones can be seen when a female mouse is exposed to the urine of a male mouse unfamiliar to her. This circumstance prevents pregnancy and is called the Bruce effect after Hilda Bruce, who first discovered it. The Bruce effect is mediated by neurons of the vomeronasal organ that activate a hypothalamic circuit that suppresses the release of pro-lactin, which in mice is necessary for pregnancy.[40] The ability of some animals to recognize the odor of specific individuals thus seems dependent upon the vomeronasal organ.

Other pheromones act as signals between animals that control the initiation of sexual behavior. There is a simple way of demonstrating this: if a male rat is placed into a T-shaped maze, with another male at the right end of the T and a female at the left end, the first male will almost always turn left toward the female, even if he cannot see or hear her. This sexual preference is dependent upon the sense of smell and is eliminated by damage to the accessory olfactory bulb or to the SDN, which receives signals from the accessory olfactory bulb.[11] Thus, it might be that a job of the SDN is to establish a preference for females in males by reacting to the pheromones secreted by females.[10]

Do humans utilize the vomeronasal organ to detect pheromones? A small vomeronasal organ does exist in the human nasal mucosa, but analysis of the human genome shows that only 4 of 200 pheromone receptor genes have survived intact during evolution. The remaining genes for human phero-mone receptors have acquired DNA alterations that have rendered them into nonfunctional "pseudogenes." Thus, it is likely that the vomeronasal organ does not meaningfully respond to odors in humans.[17] So, many of the adver-tisements for musky perfumes that are supposed to be sexually attractive may not have a foundation in reality. However, this does not necessarily mean that the sense of smell has nothing to do with sex behavior in humans, even if we have lost the special abilities of the vomeronasal organ. A number of studies have shown that if human subjects are allowed to inhale the scent of an androgen-like compound that is found in sweat, the hypothalamus is acti-vated. Moreover, the patterns of activation differ between women and men.[33]

How can we find out how much this laboratory work in rats applies to humans? It turns out that almost all of the sexual differences in central ner-vous system (CNS) structures that have been found in rodents also seem to be present in the human nervous system. While humans may have lost their sensitivity to pheromones during evolution, this does not mean that the hu-man SDN has no function. It still is conceivable that the human male SDN may "label" nonolfactory features of women as sexually desirable so that sexual preferences in humans may be under the influence of the human SDN. One study by a researcher named Simon LeVay has shown that male homosexuals have a smaller SDN than male heterosexuals, and it has been proposed that this hypothalamic structure may play a part in establishing the altered sexual preferences of human homosexual males.[10]

Studies of the brain structures that regulate sexual behavior in humans can be hard to interpret. A smaller SDN in human male homosexuals would be consistent with data showing a role for the SDN in determining the sexual preference of a male rat for female rats. However, if a diminished size of the SDN is also associated with a lower level of male sexual behavior in a rat, wouldn't that suggest that human male homosexuals would have a diminished ability for sexual arousal, just like male rats that have a smaller SDN? This is probably not true.

More recent studies in humans have used novel methods to try to test the proposal that the hypothalamus drives male sexual behavior. These studies have yielded results that generally support most of the concepts that originated from research in rats.

One of the best ways to study the involvement of a brain structure in a behavior in humans is a method called functional magnetic resonance imaging (fMRI). You probably already know what an MRI is: this imaging technique utilizes intense magnetic fields to produce electronic "slices" of various tissues. For example, if you have torn a ligament in your knee, an MRI is the optimal way to detect this, since MRIs can image soft tissues like muscles and ligaments, whereas X-rays can generally only image skeletal tissues. MRIs can also provide views of the soft tissues of the brain and allow the cortex, corpus callosum, thalamus, and hypothalamus to be distinguished from each other in pictures.

A functional MRI is even better. In this technique, the amount of oxygenated blood flowing through a given brain structure is estimated. This is possible because blood contains the iron-rich compound hemoglobin. Oxygenated hemoglobin produces a different signal in MRIs than deoxygenated hemoglobin. Thus, the right kind of MRI can detect which regions of the brain are using the most oxygen. Any brain region that is activated also takes up more oxygen from the blood. So, for example, if a person is shown lots of pictures on a screen placed within the cavity of an MRI machine, his visual cortex will become activated, and the fMRI will show an increased use of oxygen in this brain region.

This approach can be applied to find out which regions of the brain are active during sexual arousal as well. One recent study of this kind was performed by a group of researchers in Rome, Italy.[6] In this study, 18 young male volunteers were asked to lie down on a platform within the confines of the large cylinder suspended within the magnetic field of an MRI machine. Then they were asked to watch short movie clips projected onto a mirror in front of their face. Some of the movie clips were of neutral activities like sports events, while others showed a man and a woman having sexual relations. The degree of sexual arousal in each subject was also monitored by a device that measured penile volume. The study showed that erotic scenes, but not sports scenes, consistently produced both sexual arousal and an activation of the hypothalamus. I find it a bit difficult to visualize what this experiment must have been like for the volunteers, but it apparently worked, so I guess a person's interest in sex will withstand all kinds of uncomfortable and peculiar surroundings!

If the SDN is concerned with sexual preference, and if spinal neurons operate the muscles of the genital area, what specific hypothalamic neurons are responsible for penile erection and sexual arousal? Many studies have shown that the paraventricular nucleus (PVN) of the hypothalamus has a critical influence upon this aspect of male sexual behavior. Electrolytic lesions of the PVN eliminate penile erection in rats, and tiny amounts of neurotransmitters can potently stimulate penile erection when infused into the PVN. PVN neurons that make oxytocin as a neurotransmitter send axons all the way down to the lumbar spinal cord and seem responsible for all of these effects. These axons activate parasympathetic fibers that innervate the blood vessels of the penis.

DRUGS AND SEXUAL FUNCTION

A number of drugs can interfere with the function of the PVN neurons that stimulate penile erection. Morphine and other opioid narcotics, for example, block the activity of PVN neurons and are probably responsible for the symptoms of lessened sexual desire reported by people addicted to heroin or to related opium-like drugs. Also, the active component of marijuana, a molecule called a cannabinoid, has similar effects upon sexual potency. Marijuana affects the brain because it mimics the activities of another chemical (called anandamide) that is made as a neurotransmitter within the brain. Activation of cannabinoid receptors in the PVN also decreases sexual behavior in rats.[7]

One specific type of drug that is widely prescribed by physicians to treat depression also has a major inhibitory effect upon male sexual behavior. Drugs belonging to this category are known as selective serotonin reuptake inhibitors (SSRIs). These drugs affect the brain by preventing the reuptake of a neurotransmitter called serotonin back into nerve cells. Examples of these drugs include citalopram (Lexapro), fluoxetine (Prozac), and sertraline (Zoloft). These drugs cause serotonin to accumulate in the spaces between nerve cells and result in a prolongation and an enhancement of serotonin's effects. These drugs seem to be effective in reducing severe feelings of depression and have become increasingly popular: in 2007, almost 100 million prescriptions were written for these types of drugs in the United States alone. This widespread use of SSRIs has not gone without some criticism. In patients with mild depression, they do not seem to be much better than a placebo (inactive pill).[13] Also, a major problem with these drugs is that they can decrease libido and interfere with the ability of a man to undergo penile erection and ejaculation in 20% to 40% of patients. How do these unwelcome effects come about?

Research in rats has recently identified why these useful drugs can have such unfortunate "side effects" on male sexual arousal. It turns out that there is a large cluster of nerve cells in the medulla of the brain called the nucleus paragigantocellularis, which gets its unwieldy name from the fact that it is near another nucleus that has unusually large nerve cells. This cluster of neurons projects to the ejaculatory nucleus in the spinal cord and inhibits its function. The nucleus paragigantocellularis receives lots of synapses containing serotonin, which activate its nerve cells. When serotonin levels in the brain go up, as happens when a man takes an SSRI-type antidepressant drug, this nucleus goes into overdrive and suppresses the ability to undergo male sexual behavior.[27]

Antidepressant drugs have an unquestioned value in the treatment of severe depression and can potentially reduce the risks for suicide in depressed individuals. Their sexual side effects, however, are not desirable. Increasing attention to this conflict between the "good" and "bad" effects of these drugs is leading to new attempts to find a drug that would increase the effects of serotonin on brain regions that modulate mood but would leave the function of the circuits of sex behavior unaffected.[27]

AGING AND MALE SEXUAL BEHAVIOR

During aging, testosterone production shows a steady decline: between the ages of 30 and 70, blood levels of free testosterone decrease by almost 50%.[18] This decrease in testosterone causes the testosterone-sensitive nerve clusters in the spinal cord and in sympathetic ganglia to become less active, leading to erectile dysfunction and to lessened sexual activity in older men. This decrease in testosterone is puzzling. The secretion of many other hormones in the body is not nearly as impaired during aging as is the secretion of reproductive steroid hormones. Why do testosterone levels fall like this? Does aging somehow preferentially affect testosterone-producing cells?

The cells in the testes that produce testosterone are called Leydig cells. These cells adhere to the outer surfaces of seminiferous tubules that contain developing spermatozoa. Each testis contains about 25 million of these Leydig cells, and all of them energetically produce lots of testosterone every day. Most studies have shown that the numbers of these cells seem to remain stable throughout our lives. Also, research using a peculiar chemical called ethane dimethane sulfonate (EDS) has shown that Leydig cells have a remarkable potential for regeneration.

If a rat is injected with EDS, this chemical has the amazing ability to selectively kill off all the Leydig cells in the testes while causing no harm whatsoever to any of the other cells. This is because Leydig cells and Leydig cells alone have an enzyme that converts EDS into a lethal toxin. Within two to three days after an EDS injection, all the testicular Leydig cells die, and blood levels of testosterone fall into the basement. Such treated rats have little interest in sex. However, seven weeks later, unharmed stem cells in the testes "wake up" and almost miraculously start to divide and create brand new Leydig cells that completely restore levels of testosterone to normal and rejuvenate the rat.[8]

Because of this remarkable ability for regeneration, it is not likely that aging could substantially reduce Leydig cell number per se. However, it has been shown that aging does depress the ability of Leydig cells to respond to a pituitary hormone (luteinizing hormone) that stimulates them to produce testosterone. The mechanism of this aging-induced impairment of Leydig cells is not known for sure. It does seem to involve the production of harmful chemicals called free radicals in the aging testes. In rats, at least, administration of vitamin E seems to reduce these harmful effects of aging.[8]

Thus far, the main recommended treatment for this aging-associated loss of testosterone production and male sexual activity is a drug called Viagra (sildenafil). This drug does not act upon testosterone-sensitive neurons at all. Instead, it relaxes the smooth muscle in the blood vessels of the penis by preventing them from degrading cyclic GMP. This causes these vessels to become more responsive to nerve signals even in the absence of high levels of testosterone.[38] Viagra is thus widely used as a treatment for erectile dysfunction and has brought millions of dollars into the coffers of Pfizer, the drug company that first tested and marketed Viagra.

This remarkable ability of Viagra to combat erectile dysfunction is due to its effects on an enzyme called phosphodiesterase in the blood vessels of the penis, which normally degrades cyclic GMP. Viagra prevents phosphodiesterase from doing its job and keeps the blood vessels of the penis swollen and filled with blood. These welcome effects of Viagra are not, however, without some drawbacks. Viagra also has some blocking effects upon other forms of phosphodiesterase that are found in blood vessels throughout the rest of the body. So it can also dilate these blood vessels.

Dilation of blood vessels in the brain causes the headaches that can be a common side effect of Viagra. More seriously, Viagra can also dilate the blood vessels in the heart and other regions of the circulatory system. This is not normally much of a problem, but it can be a serious side effect in people who are already taking other drugs. For example, it is not uncommon that a 60-year-old man with a history of heart disease will take a drug called nitroglycerin (yes, it *is* the same compound that can be used in explosives!). This drug acts upon blood vessels in a way similar to nitric oxide and can cause dilation of the coronary arteries of the heart. This, of course, is a good thing, particularly if the circulation in the heart is not all that great, leading to symptoms of heart pain called angina. However, if the same 60-year-old man takes Viagra on the same day he has taken nitroglycerin, the two drugs will have additive effects upon blood vessels, making them dilate too much and causing a dangerous drop in blood pressure. Before these drug interactions were well understood, Viagra was involved in the sudden deaths of several hundred unsuspecting men. Fortunately, nowadays these dangers are well recognized, and labels on the bottles of Viagra specifically warn against taking the compound along with other vasodilating drugs. Millions of prescriptions for Viagra are filled every year with, now, a very good safety record.

FEMALE SEXUAL BEHAVIOR

In rats, the expression of female sexual behavior involves an arching of the back called lordosis, which permits a male to have easier access to the female's genital area. Also, in response to the female sexual hormones estrogen and progesterone, female rats make high-pitched squeaks (in the ultrasonic range that we couldn't detect in our lab unless high-speed tape recorders were used). Finally, females perform movements such as ear wiggling that seem to drive male rats wild. These stereotyped behaviors of rats can be easily analyzed to assess which areas of the brain are needed for female sexual behavior. They also led to some sophomoric humor among us graduate students. When we were in the hospital cafeteria eating lunch and happened to observe a girl flirting with a guy, we all said that some ear wiggling must be going on. This is the classy type of interaction that you can expect from prospective PhDs.

It has long been known that estrogen-sensitive neurons exist in the midportion of the hypothalamus and that lesions of this area prevent female rats from exhibiting these features of female sexual behavior. Once again, though, until recently it has been difficult to pinpoint the exact area of the hypothalamus needed for this type of behavior. A recent breakthrough has now clarified the situation.

This breakthrough originated in the lab of Dr. Keith Parker in Texas. Parker at first had little interest in the hypothalamus at all. He was interested in understanding how the adrenal glands, ovaries, and testes develop and what signals cause these organs to secrete steroid hormones like cortisol or estrogen. He discovered that a DNA-binding protein called steroidogenic factor-1 (SF-1) turned on the genes for steroid synthesis and gonad development. When mice were genetically engineered to lack this protein, they failed to develop any ovaries or adrenal glands!

Parker then asked the question of whether SF-1 might be present in any other organs besides the gonads. He was surprised to find that it was present in the brain, but only in one structure: the ventromedial nucleus of the hypothalamus (see Fig. 1-3 for a good picture of this nucleus). This seemed to suggest some mystical connection between this hypothalamic structure and reproduction. Once again, Parker genetically engineered mice to lack SF-1, and this time he examined the brain. He was astonished to see that such mice had mostly normal hypothalamic structures, but with one exception: the ventromedial nucleus had completely vanished and was replaced by a sparse collection of poorly organized neurons! Thus, the DNA-binding protein, SF-1, is required for hypothalamic neurons to organize themselves into the ventromedial nucleus, and mice that lack this protein have an abnormal hypothalamus.

When the behavior of mice lacking a ventromedial nucleus was examined, it became quite clear that they have a serious problem: female mice lacking this nucleus have severe deficits in female reproductive behavior. These experiments show that estrogen-sensitive neurons in the ventromedial nucleus are critical for the expression of female sexual behavior. It is likely that these neurons project to other neurons (in the central gray of the midbrain and in the spinal cord) that control the postures seen during sexual behavior.[22]

Do any of these studies in mice have any relevance to humans? It turns out that there is a human gene for SF-1, and that some humans inherit abnormal forms of this gene. Some patients have abnormalities in adrenal function, while others show abnormalities in ovarian function. Recent studies have shown that an abnormal SF-1 protein contributes to a condition called premature ovarian failure, in which women cease menstruating before the age of 40 and have reproductive difficulties. This condition occurs in about 1% of all women. Brain anatomy and sexual behavior in these women have not been specifically examined, but it seems likely that the symptoms produced in mice lacking SF-1 may partly occur in humans with mutations in SF-1 as well.[25]

One of the most startling early papers on the control of female sexual behavior was generated by Charles Phoenix and coworkers in 1959. They found that if pregnant guinea pigs were briefly treated with the male sex hormone, testosterone, the offspring of these guinea pigs experienced permanent changes in sexual behavior. Females no longer experienced lordosis when treated with estrogen and progesterone; in other words, they became permanently "defeminized."[41, 42] More recent studies suggest that this is because the estrogen-sensitive neurons of the ventromedial nucleus no longer respond properly to estrogen.[29]

What happens if human infants are exposed to high levels of male hormones during development? This experiment cannot be ethically performed in humans, but an "experiment of nature" may provide the answers to this question. The experiment I have in mind is a medical condition called congenital adrenal hyperplasia (CAH), which is caused by an abnormal function of an adrenal enzyme called 21-hydroxylase. Cells of the adrenal gland normally utilize this enzyme as part of a synthetic pathway for cortisol and other adrenal hormones. When the enzyme is dysfunctional, molecules meant to be transformed into cortisol become shunted into another pathway for the synthesis of male hormones like testosterone. Because of this, developing baby girls are exposed to very high levels of male hormones in utero.

Baby girls that are born with this hormonal disorder are usually promptly given other adrenal hormones like cortisol that allow them to have a normal, healthy life and which suppress the hormone-producing capacity of their own adrenal glands. Because of these treatments, such girls grow up normally and have a normal onset of ovarian function. However, while most of these CAH patients are heterosexual, they tend to have fewer sexual encounters or partners of any type than a control population and also report more interest and activity in homosexual sex.[26] So, it seems that the information gleaned from female rats exposed to testosterone before birth may be relevant to the sex lives of women as well. One thing that has not been examined is whether or not the sexually dimorphic nucleus of women with CAH is different from that of control women. An answer to this question may be long in coming, since CAH is relatively rare and it might take some time to accumulate brains donated from these patients so that hypothalamic anatomy can be studied.

AGING AND FEMALE SEXUAL BEHAVIOR

The loss of ovarian hormones that is seen at menopause seems to have adverse effects upon the ability of women to enjoy sex. It is estimated that about 52% of postmenopausal women—amounting to 15 million individuals—experience a lessened libido. One way to treat this condition is to give women pills containing different forms of estrogen and/or testosterone. Estrogen alone does have some stimulatory effects on libido in women, but if testosterone is also added to the treatment, the combination of the two hormones is much more effective than just estrogen alone. This should not be as surprising as it sounds. Normally, women have substantial amounts of testosterone circulating in their blood, though not as much as in men. Some of this testosterone is produced directly by the ovaries and adrenal glands, and some of it is converted into testosterone from estrogen by enzymes in a variety of tissues. Both of these steroid hormones seem necessary for sexual desire and activity.[23] Some caution should be employed in prescribing sex steroids to older women, however, because these hormones can also increase the risk for blood clot development and for other potentially harmful effects.

Other approaches have also been tried to remedy the decrease in sexual desire seen after menopause. Viagra, for example, seems to affect blood flow in the clitoris, just as it does in the penis. However, most studies of Viagra in women have shown no substantial effects upon sexual desire or performance, which has been a big disappointment for the drug manufacturers in search of another gold mine for their product in the nonmale half of the population.[4]

SEX, EUPHORIA, AND PAIN SENSITIVITY

One of the welcome "side effects" that can be experienced following lovemaking is a sensation of euphoria and even a diminished sensation of the aches and pains that plague us as we get older. Are these sensations real or only subjective? In fact, these welcome changes in mood following sex in humans have been detected and objectively studied in rats. Following sex, rats also display a diminished ability to react to annoying stimuli such as small electrical shocks to the tail.[15,24]

Sex-induced analgesia probably occurs because of the release of neurotransmitters called endorphins, which modulate the pain-sensitive circuits of the spinal cord. Endorphins were first discovered in the 1970s as small proteins in the nervous system that diminish sensations of pain. They were given the name "endogenous morphines" because they bound to the same receptors on pain-regulating neurons that are occupied by morphine. It turns out that the morphine produced by opium poppies has a chemical structure that is very different from endorphins, but morphine nevertheless can act within the brain because its three-dimensional shape is still similar to the three-dimensional shape of endorphins and thus can bind to the same receptors.

Plants produce these kinds of chemicals to protect themselves: when a cow eats an opium poppy, it will feel disoriented and strange and will likely avoid eating the poppy again. These plant-protecting chemicals have found a place in medicine because they mimic the effects of endorphins. Another artificial chemical called naloxone can prevent the effects of both morphine and endorphins by covering over and inactivating the endorphin receptors. Naloxone can also block sex-induced analgesia, so endorphins must be involved in this phenomenon.[15]

TRANSSEXUALITY AND THE HYPOTHALAMUS

Small numbers of individuals in our population express the opinion that they feel like they should inhabit the body of the opposite gender (e.g., a man trapped in a woman's body). These feelings are relatively rare (about 1 in 30,000 people express these feelings) but are powerful and potentially very disruptive of a happy life. Some studies have shown that abnormalities in the sequence of estrogen receptors are associated with an increased risk for transsexuality. Also, about 1% of women with congenital adrenal hyperplasia develop feelings of transsexuality, a rate many times higher than that of the general population, suggesting that exposure to hormones during development could influence the development of transsexuality.[3] Finally, abnormalities in the anatomy of the sexually dimorphic nucleus (and also in a nearby structure, the bed nucleus of the stria terminalis) have been found in transsexuals. It seems likely that syndromes of gender confusion are *not* due to peculiarities in upbringing or social environment but instead are related to processes of masculinization of the hypothalamus during development.[3]

SEXUAL FETISHES

The ability of nonsexual objects to have erotic connotations for some people is kind of a puzzle for most of us. For example, why do some people find feet sexy? Why not elbows or knees or noses? What is so special about feet?

There is as yet no certain way to answer these questions. It is certainly true that some components of our sex lives are not "hardwired" into the brain and that sexual behavior cannot be simply boiled down into a series of simple reflexes dominated by the hypothalamus. Human beings are more complex than mice or rats, and they certainly can learn to associate odd types of stimuli with sexual pleasure in a way that other animals do not.

There is one hint about the organization of the cortex of the brain that may shed light on sex and feet. The portion of the cortex that directly responds to tactile (touch) stimuli is called the somatosensory cortex. It forms a strip that begins on the surface of the brain just beneath our ears and proceeds upward to wrap around the top of the brain and down into the midline where the cortex forms an inner (mesial) layer. All of the sensations from the body are organized to be detected by specific regions along this strip. Sensations from the face and fingers occupy most of the somatosensory cortex. Sensations from the feet are received from a small mesial portion of the cortex that is not visible from the outside of the brain. Curiously, sensations from the genitalia are received by a small portion of the cortex that is just anterior to the sensory area for the feet.[1] Is it possible that some kind of overlap between these two closely adjacent cortical areas could be involved in generating the overlapping interest in sex and feet that is found in some people? Electrophysiological studies would be needed to answer this question.

WHAT IS SEX GOOD FOR?

Sex is good for all of us on a personal level, but at the level of our species, it is supposed to be beneficial because it promotes genetic diversity among us. What does that mean, exactly, and why would this be a good thing?

In most of the cells of all human beings, 23 pairs of chromosomes are found within the cell nucleus. For each pair, one of the chromosomes was inherited from the mother, and the other chromosome was inherited from the father (e.g., chromosome 1 comes in two copies, one from the mother and one from the father). Normally, most cells divide via mitosis. In mitosis, each chromosome is copied to make two chromatids joined together at a region called the centromere. Then, all the chromatids are moved apart into two separate daughter cells. The chromatids now are once again called chromosomes and are all identical to those of the original cell.

A very different thing happens during meiosis. Prior to meiosis, each chromosome is once again copied to make two linked chromatids. However, the two chromatids of paternal chromosome 1 and the two chromatids of maternal chromosome 1 become temporarily bound together in a process called synapsis. During synapsis, parts of the DNA chain of each chromosome become physically cut and then spliced back together with other chains of DNA. This bizarre event is called crossing over.

Meiosis permits germ cells to be created that have only half the normal number of chromosomes. This is important, because the sperm and egg cells will later fuse together to form the cells of a baby, and a baby's cells cannot have twice the normal number of chromosomes. Also, the baby now will have chromosomes that are all mixed together and thus have combinations of genes that are slightly different from either the mother or the father.[44]

So what? Is it so important that a baby have slightly different versions of genes than his parents? Aside from minor differences in size or shape, all humans look relatively similar, so what is the fuss about sex? Why is it so important?

The value of genetic diversity can be illustrated by looking at populations of animals that have lost this diversity. One such population that has attracted a lot of attention lately is the Tasmanian devil.

These fierce little creatures inhabit an island off the coast of Australia. They were once very abundant there, but in the 1920s they were regarded as predatory pests and a substantial effort was made to eradicate them from the island. Since then, this policy has been reversed and breeding programs have been put in place in an attempt to restore the diminishing numbers of these marsupial carnivores. These programs, however, have encountered a serious problem.

Because the original population of devils had been decimated by eradication efforts, most of the surviving devils have been derived from a very small founding population that was inbred to produce more devils. As a result, most Tasmanian devils are closely related to each other and lack much genetic diversity. Because of this, they are unfortunately able to transmit a vicious form of cancer simply by biting each other!

It turns out that, some time ago, one devil developed a type of cancer of the jaw due to some kind of mutation in a cell that became a tumor. Devils frequently tussle and bite one another, and it is easy for a cancer cell to jump off of one devil into the mouth of a second devil. Normally, this would be no problem because the immune system of the second devil would recognize the cancer cell as a foreign invader and promptly destroy it. This does not happen today, however, because the genes of most devils are now so similar to each other that their immune systems do not react against each other's cells. This allows a tumor cell to jump from one creature to another and gradually destroy thousands of animals.

The spread of cancer within Tasmanian devils shows how important a certain amount of genetic diversity even within a single species can be. Sex, therefore, is more than a passing pleasure and is important for the survival of most species.[35]

THE HYPOTHALAMUS AND LOVE

Mating is not the only component of reproduction but in fact is only the beginning. It allows only for the conception of offspring. In order for babies of warm-blooded animals like birds and mammals to survive, their parents must also be stimulated to love and care for them. The hypothalamus has a commanding role in generating these behaviors as well.

Maternal behavior in rats involves picking up the rat babies (pups), building nests out of bedding material, and suckling the pups. Rat pups are not particularly cute, at least to my eyes: they are hairless and blind (their eyelids have not yet opened) and look a lot like little pink erasers. Nevertheless, they are very appealing to female rats, but only in the right circumstances. A young virgin female rat will not spontaneously exhibit maternal behaviors in the presence of pups and will avoid or even attack them. However, if the same rat is injected with estrogen and then given intracranial infusions of oxytocin, all of these maternal behaviors appear.

It turns out that the process of giving birth and nursing pups normally stimulates the release of oxytocin from nerve cells in the supraoptic and paraventricular nuclei of the hypothalamus (see Fig. 2-1 in the previous chapter for a view of these nuclei). Oxytocin acts on mammary glands to stimulate the release of milk. But the effects of oxytocin do not stop there. Oxytocin is also carried from the hypothalamus to many other parts of the brain to affect behavior. One of these brain regions is the amygdala, a part of the temporal lobe of the brain that seems to add an emotional "color" to sensory stimuli processed by the brain. These neural effects of oxytocin are the basis for maternal behaviors in rodents.

A hypothalamic peptide with a structure closely related to oxytocin is vasopressin, which is also produced by neurons of the paraventricular and supraoptic nuclei and is sent from the hypothalamus to other regions of the brain. This peptide has also been shown to affect behavior. A striking example of the effects of vasopressin can be seen in studies of a small burrowing mammal called a vole. A number of species of voles exist, and these species differ in their social structures and behaviors. Prairie voles live in large burrows with extended families and mate loyally for life with a single spouse. Meadow voles, in contrast, live solitary lives and gaily mate with any partner who becomes available. Why do these closely related species behave so differently?

When the brains of these two species were examined under the microscope, it was found that the patterns of binding of vasopressin to brain structures were very different between the two species. Also, if vasopressin is infused into the brains of these animals, it stimulates bonding between male and female and makes the animals more sociable. It seems likely that inherited alterations in the function and distribution of the vasopressin receptor in the brain account for these social differences between prairie voles and meadow voles.

Does vasopressin or oxytocin also affect human behavior? This question is a little more difficult to answer in humans than in voles. However, inherited differences in the structures of the vasopressin receptor in humans do seem to correlate with how altruistic a person will behave when playing a video game during an experiment.[20] Also, nasal sprays containing oxytocin seem to increase measures of empathy in human subjects; these effects are not seen in patients with damage to the amygdala.[19] So it may be that the effects of vasopressin and oxytocin seen in rodents are broadly applicable to humans.

THE HYPOTHALAMUS, ANGER, AND ANXIETY

The benign types of emotions and activities just reviewed are not the only behaviors influenced by the hypothalamus. Anxiety and anger also seem to be subject to the control of this brain region.

The influence of the hypothalamus on emotions became apparent during the early years of study of lesions that cause increased appetite and obesity. An altered appetite was not, however, the only effect of such lesions. I can remember from my graduate student days how easy it was to identify which rats in a rat room had received VMH lesions. When I entered the rat room and softly clapped my hands, most of the rats would simply turn around in their cages, look curiously out through their tiny metal bars, and quietly sniff at me. VMH-lesioned rats, however, would jump dramatically and push against their cage walls, virtually snarling at you. If a researcher wanted to pick up a VMH-lesioned rat, he would have to put on special chain-mesh gloves to protect himself; otherwise the rat would try to eat his fingers! This type of behavior was long ago termed "sham rage," but it sure looked like real rage to me! Also, these lesioned rats were strikingly different from ordinary lab rats. Once a rat has become accustomed to being handled, it can become quite docile and even make an excellent pet. Many times I have had a rat climb up my lab coat, sit on my shoulder, and lick the salt off my ears. So the emotional state of a VMH-lesioned rat is really remarkable.

We now know, from studies of animals lacking a ventromedial nucleus, that much of this behavior stems from an apparent increase in anxiety that occurs after hypothalamic damage.[21] This information may also apply to humans. Occasionally, human patients develop a specialized type of brain tumor called a hypothalamic hamartoma. Initially small, growing tumors can provoke peculiar episodes of uncontrollable smiling or giggling. A patient will start laughing for no apparent reason, as if enjoying some secret joke. As the tumors progress, they can provoke epileptic seizures and episodes of uncontrollable rage. Some people affected with hypothalamic tumors will get into fights or, with little provocation, start pacing around with clenched fists and begin to verbally abuse others. This behavior can progress to assaults on others and lead to a sad record of delinquency and troublemaking.[32] So the way we view the world emotionally can be dramatically affected by the hypothalamus.

Sex differences also come into play when the control of aggressiveness is examined. Male rats are more aggressive than female rats, and much of this increased aggressiveness seems to arise from exposure to testosterone during development. Thus, the sexual dimorphisms in the hypothalamus may be related to more than just the expression of sexual behavior.

Some recent direct evidence for this in humans comes from studies of the testosterone (androgen) receptor protein in men. The structure of this protein shows surprisingly wide differences between individuals, mainly because the DNA coding for this protein can have variable numbers of repeats of DNA bases in them. Proteins generated from DNA having a smaller number of repeats of these bases seem to be more responsive to testosterone and thus cause the cells containing these receptors to react more vigorously to this hormone. A number of scientists have studied these receptors in populations of normal men and also in populations of men incarcerated in prison for violent or sexual crimes. The data available thus far show that if a man inherits a stronger type of androgen receptor, that man will tend to be more impulsive and aggressive and seems to be at higher risk for committing violent crime.[40]

So far in this book I have tried to argue that variations in the receptors for sex steroids in the brain can be associated with alterations in a number of behaviors, as in anorexia nervosa (chapter 1) and aggressive behaviors (above). How could apparently minor changes in these receptor proteins have such drastic effects on hypothalamic neurons? How do sex steroids affect the function of nerve cells?

Receptors for sex steroids can dramatically affect many aspects of cell function because they directly affect the functions of genes within chromosomes. Sex steroids are basically modified forms of a lipid molecule called cholesterol. When a sex steroid is secreted from the ovaries or testes into the bloodstream, it is carried along in the blood by being attached to proteins; otherwise, as a fat (lipid) molecule, it wouldn't dissolve in watery blood at all. If the steroid becomes close to a cell, it pops off of its carrier protein and first appears to interact with proteins of the cell membrane. Occasionally, effects of steroids upon neuronal cell membranes alone are sufficient to rapidly activate a nerve cell, but in general, to change neuronal function, a steroid molecule is usually moved to the cell nucleus. Once inside the nucleus, it binds to steroid receptors, which in turn now can bind to specific regions of DNA and affect the transcription of DNA into messenger RNA. New messenger RNAs then travel out of the nucleus into the cytoplasm, where they are utilized to make new proteins. Estrogen itself seems to be able to change the transcription of at least 400 different genes to make 400 different proteins.[37] The details of how steroid receptors bind to DNA and affect gene function are actually rather complicated and are probably better subjects for more technical books that I would encourage interested readers to peek into.[37,45]

SUMMARY

All of these kinds of information show that it shouldn't be too surprising that any alteration in the ability of sex steroid receptors to modify nerve cells could cause a wide variety of changes in the brain. A basic core of neural tissue that responds to steroids resides in the hypothalamus and projects to lower regions of the central nervous system to regulate behaviors such as male sexual behavior, female sexual behavior, love and affection, and aggression. This is not to say that human beings are hapless robots doomed to carry out the simple commands issued by the hypothalamus. Humans and other animals have a great capacity for learning, and there is no doubt that learned cues processed by the cerebral cortex and other brain regions also have an important influence on the types of behaviors we experience. When we sense a welcoming smile, a subtle shift in body language, a frown, or more overt gestures from others, we analyze these signs in very complex ways to guide our actions. But the basic machinery that translates our feelings into actions definitely must pass through the hypothalamus before changes in behavior occur.

· 4 ·

The Hypothalamus and
the Control of Sleep

$\mathcal{F}$or most of this book, I've tried to show you how really overwhelming the influence of the hypothalamus can be in our lives. The hypothalamus is a very small structure in comparison to the remainder of the brain, but urgent signals from it can overcome conflicting impulses generated by much larger and more complex brain structures like the cerebral cortex. When the hypothalamus detects a fall in body water, we become obsessed with finding a drink of water. When the hypothalamus determines that we don't have enough calories and are starving, we think of nothing but food and tend to forget about what might be on television or what books or music we might be missing. Hypothalamic stimuli for sexual activity are perhaps not quite so overwhelming as other signals, but they certainly can have strong influences on our lives. One final signal from the hypothalamus that can override other considerations is the need for sleep. Very sleepy people find it very difficult to think about anything except getting into bed and catching some z's. Once again, the hypothalamus plays a major role in creating this fundamental drive.

Sleep. Sleep is a bizarre behavior that occupies almost one-third of our lives but which is still poorly understood. During sleep, we lose consciousness of our surroundings and often experience peculiar hallucinations that we call dreams. During these dreams, we become partly paralyzed and cannot move our limbs; presumably this prevents the awful injuries that could result from dreaming that we are in an Olympic marathon and acting upon this dream while asleep! While we all take these peculiar phenomena for granted as normal parts of life, when you think about it, sleep is really strange. The exact functions of sleeping and dreaming are not known for certain, even though sleep has been studied by scientists for at least the last hundred years.

In the 19th century, numerous physiologists tried in vain to adequately explain sleep. One popular theory was that stress and strain of muscles caused them to release substances like lactic acid into the bloodstream which absorbed oxygen and depressed the normal function of the brain. Once the muscles had recovered from their stresses, they would stop releasing these mysterious substances and a person would wake up again. These kinds of theories at least acknowledged that prolonged periods of activity tend to make us sleepy, but they really didn't identify the true mechanisms that caused sleep.

The first real breakthrough in sleep research came in 1929 when a German physician named Hans Berger was able to create a device that could detect electrical currents produced by the brain. He attached small silver plates to the scalps of his subjects and connected them via wires to a device called a galvanometer. This setup was so primitive that the "noise" in the system almost drowned out the data Berger was able to acquire, but in the final analysis it did allow him to detect waves of electrical activity that repeated 10 times a second. These waves were subsequently called alpha waves of the electroencephalogram (EEG). He also reported that the EEGs of sleeping subjects differed from those recorded in patients who were alert and awake.[3]

The next breakthrough in sleep research occurred in 1953 in the lab of Nathaniel Kleitman, a Russian émigré who had become a professor of physiology at the University of Chicago. Dr. Kleitman and a graduate student named Eugene Aserinsky were studying a curious thing: when the sleep subjects they were studying settled into a steady sleep, their eyeballs rolled around beneath their eyelids as if they were looking around. One of these patients seemed very agitated during these eye movements, and when he was awoken, he described the vivid dream he was having.

As Kleitman and Aserinsky continued to study their subjects, they found that this rapid eye movement phase of sleep always occurred during episodes of dreaming and were always accompanied by a sudden increase in the frequency of the electrical rhythms of the EEG. They concluded that sleep involved changes in the activity of the cerebral cortex and that sleep could be divided into a number of phases. The two main kinds of sleep are rapid eye movement (REM) sleep, which occurs when we are dreaming and which is accompanied by eye movements that track the progress of our dreams, and non-REM sleep. Non-REM sleep is deeper and accounts for about 50% of total sleep time. REM sleep accounts for about 20% of sleep time, with intermediate stages of sleep making up the rest of our time in bed. During all of these sleep stages, recordings of electrical impulses propagated from the brain through the scalp show distinctive changes that are characteristic for each sleep stage.[3]

Sleep does appear vital for health and life. A rat can be prevented from sleeping by placing it in a cage with a plastic disc forming the floor. When signs of sleepiness in the rat are detected, the disc can be rotated, and the rat must walk a few steps to avoid falling out of the cage into a pan of water. If this is done for 12 days or so, the rat will die, even though it had normal access to food and water. Somehow, sleep is necessary for the maintenance of normal metabolism and resistance to infection.[4]

For some time, sleep was simply regarded as a period when most brain regions become less active and we lose consciousness. Now we know that this concept is completely wrong: sleep is actively imposed upon the brain by a number of critical brain structures, and the brain is by no means inactive during sleep. One of the brain regions most critical for the control of sleep is the hypothalamus.

One early indication of the importance of the hypothalamus for sleep came from studies conducted in Vienna by a Romanian neurologist named Constantin von Economo in the 1920s. Von Economo was interested in a terrible disease called encephalitis lethargica that was sweeping through Europe and North America in an epidemic that affected thousands of people.[28] It was believed to be some form of influenza, perhaps caused by some type of avian (bird) virus.

People who came down with severe forms of this disease suffered brain damage in various parts of the brain. Initially, patients experienced a range of symptoms, including either excessive sleepiness or severe insomnia (less than two hours of sleep a night). Many people were able to recover from these initial symptoms, but one or more decades later, they gradually acquired symptoms of severe Parkinson's disease. This disease produces extreme difficulties in walking or standing and is often accompanied by tremors of the hands, a masklike facial expression, and difficulties in communicating with others. In these patients, the encephalitis had damaged a specific part of the brain called the substantia nigra, which contains movement-regulating neurons that use a chemical called dopamine as a neurotransmitter. Many of the nerve cells in this part of the brain died as a result of the infection, but the remaining neurons had enough influence to prevent any symptoms. Gradually, though, even these remaining neurons began to die as part of aging, and then the full-blown array of symptoms of Parkinson's disease appeared.

In many cases, the symptoms could be miraculously overcome by administering a precursor to dopamine called l-dopa that was able to cross the blood-brain barrier and restore the function of dopamine-containing brain circuits. The plight and partial recoveries of these patients were movingly described in a book by a neurologist, Oliver Sacks; his story was later made into a movie called *Awakenings*).[22]

Other subsets of patients with encephalitis did not develop Parkinson's disease but instead acquired severe abnormalities in sleep, including either insomnia or prolonged sleepiness. Von Economo studied the brains of these patients and found that damage to tissue bordering the third ventricle—in other words, the anterior or posterior hypothalamus—was most clearly associated with disturbances in sleep. Von Economo was nominated for the Nobel Prize for his work three times, and although he never received the award, he nevertheless enjoyed widespread acclaim for his discoveries before he died of heart disease at the age of 55.

SLEEP AND THE ANTERIOR HYPOTHALAMUS

The pioneering studies of von Economo suggested that sleep-promoting nerve cells may reside in the anterior hypothalamus and wakefulness-promoting nerve cells might be in the posterior hypothalamus. Effects of electrolytic lesions in the anterior or posterior hypothalamus in cats and rats confirmed the general truth of von Economo's findings in humans. However, which precise neurons among the millions of cells in the hypothalamus were the most critical for controlling sleep? For many years, this was not an easy question to answer.

One approach to the problem would be to see which nerve cells of the anterior hypothalamus started to fire more rapidly at the onset of sleep. Technically, however, this is difficult and would require implanting dozens of tiny electrodes into the tiny hypothalamus to see which cells began to be electrically active when a rat went to sleep. In 1996, an eminent neurophysiologist at Harvard named Clifford Saper tried another approach. This approach, first used in 1985 by Tom Curran at the Salk Institute in San Diego, involved identifying activated neurons without the need for studying their electrical properties. It turns out that an activated neuron steps up its production of a DNA-binding protein called c-fos, which accumulates in the cell nucleus about 30 minutes after the cell starts to fire rapidly. Thus, to see which nerve cells fire rapidly to stimulate the onset of sleep, Saper rapidly gave rats an overdose of anesthesia about an hour after they started sleeping. Then he removed their brains and stained sections of the hypothalamus with antibodies to c-fos (Fig. 4-1).

The results of Saper's experiments clearly showed that a small population of nerve cells in an anterior region called the ventrolateral preoptic area (VLPO) of the hypothalamus showed dramatic increases in activation during sleep. Damage to these nerve cells in rats causes a 50% reduction in the time a rat spends sleeping. In rats, the cells themselves were found to utilize a

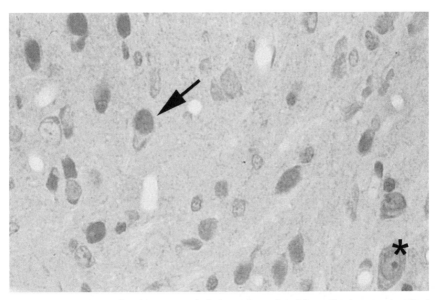

Figure 4.1.   Neurons that have recently been activated and have fired many electrical discharges down their axons possess cell nuclei that are intensely stained for the c-fos protein (arrow). Resting, inactive neurons*, in contrast, fail to stain for c-fos.

small peptide called galanin as a neurotransmitter; later, a very similar cluster of galanin-using neurons was found in the human hypothalamus, showing that an analogous population of neurons probably controls sleep in humans.[23]

How can this tiny collection of anterior hypothalamic nerve cells cause the drastic changes in overall brain function that occur during sleep? They do this by projecting to a number of other brain regions that activate the brain and cause wakefulness. One such wakefulness-producing brain structure is another small nucleus located in the posterior hypothalamus, just behind the arcuate nucleus. Neurons in this nucleus—the tuberomammillary nucleus— are the only nerve cells in the brain that utilize a chemical called histamine as a neurotransmitter. This nucleus projects to many areas of the cortex and promotes a state of wakefulness. When it is "turned off" by neurons in the VLPO, sleep follows immediately.

Activation of the anterior hypothalamus is not the only way to "turn off" histamine-containing circuits that cause wakefulness. Another way is by simply taking a pill called an antihistamine.

Antihistamine drugs (Benadryl is a good example) were initially created to combat allergies such as the sneezing and runny eyes we get when exposed to foreign proteins like pollens in the spring. These pollens provoke cells in connective tissue called mast cells to release massive amounts of histamine during an immune response against the pollen. When histamine binds to the cells of blood vessels, it causes the vessels to become leaky, so we develop a runny nose and red eyes due to leakage of blood vessels in the nasal cavity and in the conjunctiva covering our eyeballs. Antihistamines can prevent this by blocking the histamine receptors on these blood vessels. Also, if the antihistamine drug has a structure that allows it to pass through the blood-brain barrier, it blocks the effects of histamine in the brain and can cause severe sleepiness. On the rare occasions that I have taken Benadryl for an allergy, I have been overwhelmed by sleepiness and have had to retreat to bed! A current challenge for the drug industry is to find an antihistamine that provokes sleepiness (to treat insomnia) that doesn't also act on blood vessels in the rest of the body.[17]

The sleep-inducing neurons of the preoptic area are not only involved in normal sleep, but they also seem to be needed for the artificial sleep produced by anesthetic drugs. One example is a drug called propofol, which can rapidly induce sleep when injected intravenously. This powerful drug is in fact rather dangerous and was suspected in causing the death of the singer Michael Jackson. Propofol acts by activating the sleep-inducing neurons of the preoptic area.[8]

Narcolepsy. Activation of the sleep-inducing neurons of the preoptic area is one way to make a person sleepy, but it is not the only way. Another pathway to sleep is illustrated by a rare and interesting disorder called narcolepsy. Narcolepsy affects about 0.04% of the population (about 100,000 Americans). People suffering from narcolepsy have a number of troubling symptoms. They tend to fall asleep easily, particularly during the daytime when everybody else is awake. Also, they experience a symptom called cataplexy. During cataplexy, strong emotions can provoke a loss of muscle tone that causes a person to helplessly collapse even though he is fully conscious and upset by falling down.[26] If a person with narcolepsy were to walk into a surprise birthday party, he would likely fall down and go to sleep when his hosts yelled "surprise!"

What could cause such an odd disorder? A small number of people with narcolepsy appear to have inherited it from their parents, but most patients are of the so-called "sporadic" type that have no evidence of any inherited mutation in any gene. One researcher who strove to understand this disorder was Dr. William Dement, who is a major sleep researcher at Stanford University. As a graduate student, Dr. Dement had come in on the ground floor of sleep research because he had been trained by Kleitman, who had originally identified REM sleep in his lab at the University of Chicago.

One day, in 1972, Dr. Dement was showing films of people with narcolepsy to participants at a medical conference in San Francisco. A man in the audience later came forward and told Dr. Dement that he had films of a dog he had once owned that had symptoms identical to those of human narcoleptics. When the dog became excited, as when given a dish of his favorite food, the dog would fall over asleep! Dr. Dement showed this film to another conference in New York. This time, another man came forward and said he had a poodle named Monique with the same symptoms! This made it clear that dogs might be utilized in efforts to understand narcolepsy.[11]

Ultimately, a collection of narcoleptic Doberman dogs was established at Stanford so that a gene for narcolepsy could be identified by crossbreeding and genetic analysis. In 1999, after a period of 10 years, the gene for narcolepsy in dogs was finally identified. It turned out to be a gene for a receptor protein on the surfaces of neurons. This receptor protein binds a specific hypothalamic neurotransmitter called orexin.

Orexin-containing neurons are found exclusively in the lateral hypothalamus and project to the brain stem, cortex, and thalamus to stimulate wakefulness (Fig. 4-2). They also have a moderately stimulating effect upon appetite (see chapter 1). So, an inactive receptor for orexin seems to be the reason why dogs develop narcolepsy. However, what about humans?

In more than 90% of narcoleptic humans, there is not a mutation in the orexin receptor, and in fact there seems to be no direct genetic explanation for the disease. Instead, human narcoleptics have been found to have drastically decreased amounts of orexin in their brains and cerebrospinal fluid. There is a simple explanation for these findings: orexin-containing nerve cells mysteriously begin to die off during the third or fourth decade of life! Nearby neurons that don't produce orexin seem unaffected. As orexin-containing neurons die, the ability to sustain wakefulness diminishes until a full syndrome of narcolepsy appears.[20, 26]

What could be the cause of this dramatic death of only these specific nerve cells? It is basically unknown. Some form of attack by the immune system upon just these neurons is suspected, but even this hypothesis is difficult to understand, since brain tissue is protected by the blood-brain barrier. This barrier, formed by tightly sealed capillaries, should prevent cells and proteins of the immune system from entering the brain to damage orexin-containing neurons. So what is happening?

At this point, no one knows for sure how human narcolepsy comes about. One possible explanation may stem from the fact that unusually large amounts of orexin-containing axons terminate in the bottom of the hypothalamus, where the blood-brain barrier is unusually leaky.[33] Perhaps some sort of circulating toxic molecule could enter this leaky area, damage the nerve endings, and eventually cause the dysfunction and death of the nerve cells that produce these endings.

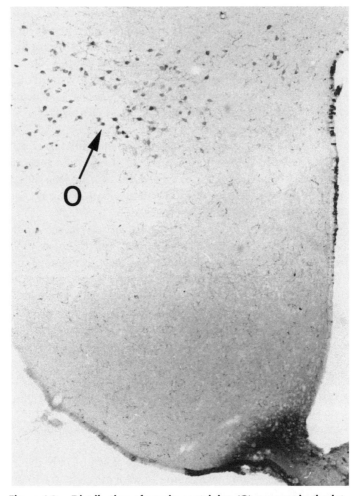

Figure 4.2.   Distribution of orexin-containing (O) neurons in the lateral hypothalamus, identified by immunocytochemistry (the use of an antibody that binds only to orexin as a stain). These cells send projections (axons) to many parts of the brain to regulate wakefulness. They also send many projections to the arcuate nucleus that also stain darkly at the bottom of this picture.

Mice can be genetically altered by using a molecular biology technique to produce orexin-containing cells that self-destruct. These so-called orexin knockout mice show abnormalities in sleep that resemble those of human narcoleptics. The total amount of time per day that these mice sleep is not different from that of control mice (about 53% of 24 hours are spent in sleeping). However, instead of having about 150 short bouts of sleeping per day, orexin knockout mice have 230 bouts of sleeping that are even shorter than normal. These periods of sleep disrupt their normal behaviors. This is probably one reason why these mice develop moderate obesity and related metabolic abnormalities.[36]

Hypothalamic neurons are not the only brain cells that regulate sleep. Other clusters of neurons in the medulla, pons, and thalamus make their own contributions to sleep and wakefulness. However, hypothalamic cells clearly have an overwhelming influence upon these other brain structures that is indispensible for normal sleep.

THE TIMING OF SLEEP AND THE SUPRACHIASMATIC NUCLEUS

One obvious thing about sleep is that most humans sleep during nighttime hours. This isn't true of many other mammals; rats, mice, possums, raccoons, and many other animals prefer to sleep during the daytime and forage for food at night, when fewer dangers from predators present themselves. If sleep is possible for many animals even during the bright light of day, what causes us to sleep mainly at night?

This timing of sleep represents only one of many functions of the human body that show 24-hour variations in activity. These variations are called circadian rhythms (from *circa* = about, and *diem* = day, in Latin). In addition to sleep, other physiological parameters (body temperature, blood levels of hormones such as cortisol, appetite, and the gut motility that produces bowel movements) show peaks during daytime hours and would greatly upset our physiology if they were delayed into the night.

These circadian rhythms are *not* strictly dependent upon the hour of the day. This can be demonstrated easily: if a rat (or a person) is confined within a room having only constant, dim illumination, the rat will still show 24-hour circadian rhythms in sleep, activity, water drinking, and so forth. However, gradually, these circadian rhythms will slowly go out of phase with the actual time of the day, so that after several weeks, a person or a rat in a dim cave or underground shelter will start to sleep in the afternoon and wake up at midnight. This type of shift in timing is called a free-running cycle. However, if a person or a rat is exposed to even a brief 15-minute burst of bright light, this free-running cycle will "reset" itself so that sleeping activity returns to a more normal timing.

What is responsible for the timing of all these activities? It turns out that the hypothalamus possesses its own internal "clock" in the form of the 10,000 or so neurons that make up the suprachiasmatic nucleus (SCN) (Fig. 3-1). These neurons show 24-hour cycles in activity. This can be demonstrated even in a brain slice that is maintained in culture in a dish.

To show these 24-hour changes in activity in the SCN, researchers genetically engineered the SCN nerve cells of a mouse to turn on an artificial gene for firefly luciferase at the same time it turned on the genes for its neurotransmitters. Firefly luciferase is a protein that allows fireflies to "glow in the dark." Later, researchers overdosed the mouse with anesthetic, removed the brain, and quickly sliced it into thin slices of living tissue that could be maintained within a tissue bath containing oxygen and nutrients. It is possible, in these slices, to distinguish the location of the SCN at the bottom of the hypothalamus. As SCN nerve cells recovered from the trauma of being removed from the skull, they began to reassert their normal properties. They began to synthesize neurotransmitters and other proteins needed for activity, and they also turned on the gene for firefly luciferase and began to glow under the microscope. This glowing activity of each cell occurred about once every 24 hours, in step with 24-hour changes in other activities of these cells.[31]

It is remarkable that these neurons, unlike any others in the brain, can independently produce 24-hour variations in activity even when separated from any sensory input that could tell them what time it is. How do they do it?

Two types of proteins are key players in running the internal clock of the SCN. One protein, called Clock, is a DNA-binding protein that stimulates the transcription of at least 127 genes on chromosomal DNA.[15] The proteins coded for by these genes keep the nerve cell in a high state of activity and force it to produce the neurotransmitters needed to communicate with other nerve cells.

The other proteins involved in running the clock are called period proteins (Per1, Per2, and Per3). When the genes for period proteins are activated by the Clock protein, molecules of messenger RNA are sent from the nucleus into the neuronal cytoplasm, where they are translated into, say, the Per1 protein. It takes quite a while (about 24 hours) for quantities of Per1 to reach a high level within the cytoplasm. When they do reach this level, two Per1 proteins bind to each other and then head into the cell nucleus. Once inside the cell nucleus, they bind to the Clock protein and prevent it from stimulating the activity of the nerve cell. The nerve cell quiets down until, gradually, all of the Per proteins are broken down and dissipate. Then the Clock protein is free once again to stimulate cell activity for another 24 hours. This cycle repeats over and over.[13]

This simple system explains how the SCN clock functions. It is not quite accurate enough to produce a true 24-hour rhythm in the absence of any sensory cues about the time of day, so when a person is placed in a dark room, the cycle goes astray. However, in a normal environment, information is continually fed into the SCN from nerve fibers carried in the optic nerves, just below the SCN. These fibers keep the SCN informed about whether it is nighttime or daytime.

One of the surprising things about the SCN is that, while well studied in rats and monkeys, it had been virtually ignored by scientists studying the human brain. The very existence of the SCN in the human hypothalamus was not even confirmed until 1980![9] In fact, there is a very good reason for this surprising ignorance about the human hypothalamus: it is often very difficult to obtain good samples of the human brain. When a scientist wants to study the rat brain, he simply gives an overdose of anesthetic to a rat and then rapidly perfuses its body with formalin so that even small brain structures are perfectly preserved and easily removed from the skull. This option is just not available to scientists who want to study the human brain: even the most scientifically oriented patient is not likely to allow someone to perfuse him just before death! Consequently, lots of the tissue samples of the human hypothalamus I have examined were obtained long after the death of the donor and just are not in good enough shape to show fine details of nerve cells. Also, frequently the bottom of the hypothalamus becomes torn or damaged during the process of brain removal from the skull. This is just one example of how the basic neuroanatomy of the human brain is even now not completely understood.

Now that the presence of the SCN in the human brain is known, it is easier to explain some medical conditions. For example, abnormalities in period proteins of the SCN can throw our internal clocks out of whack. One example is a rare inherited disorder called familial advanced sleep phase syndrome. Patients suffering from this syndrome tend to fall asleep much earlier than most of us (7:30 PM or so) and then sleep for eight hours until waking up at 4:30 AM! This disorder, while not terribly harmful, does keep these patients a bit socially isolated from the rest of us. It is caused by a mutation in an enzyme that adds a phosphate molecule to the Per2 protein. This disturbs the passage of Per2 into the nucleus and causes an abnormality in the SCN clock.[27]

A more common abnormality in the SCN clock occurs when we fly on a plane to a distant location. When we arrive at our destination, the local time may be four or five hours out of synchrony with the time zone we originated in. This confuses the SCN clock and causes the symptoms of "jet lag": sleepiness and irritability. These symptoms tend to disappear within a few days as the SCN clock resets itself to local time.

OTHER SLEEP DISORDERS

A much more common sleep-related disorder than narcolepsy, called sleep apnea, may affect as many as 1 out of 15 Americans. In this condition, the airway passages in the throat partly collapse during sleeping, causing symptoms such as loud snoring or, even worse, completely obstructed breathing that may last for a minute or so until decreasing levels of blood oxygen prompt an emergency gasping for breath.[34] This abnormal pattern of sleeping interferes with getting a good night's sleep and often causes daytime sleepiness in affected people. The causes of sleep apnea are uncertain, since people with this disorder breathe completely normally while awake. Many people with sleep apnea are more obese than normal. It may be that larger amounts of fat around the neck contribute to airway collapse while sleeping, or, on the other hand, the metabolic abnormalities caused by sleep apnea may themselves contribute to the acquisition of excess body fat.

One possible explanation centers around orexin-containing neurons of the hypothalamus. I contributed to a study of this possibility with colleagues in the Department of Physiology here at Howard University. We found that orexin-containing neurons project their axons to nerve cells in the medulla of the brain and spinal cord that regulate breathing. Even more important, when a solution containing orexin is dripped onto these respiration-regulating nerve cells, their activity is increased and breathing is augmented. Perhaps an abnormality in orexin-containing neurons contributes to sleep apnea as well as to narcolepsy.[35]

Another much rarer sleep disorder has even more puzzling symptoms. It was first identified in Europe by two different physicians—Willi Kleine and Max Levin—during the 1920s. The main symptoms of this strange disorder are attacks of sleepiness that last as long as a week and keep the patient in bed and somnolent for most of that time. When a Kleine-Levin sufferer periodically wakes up, he appears to feel extremely hungry and rushes to consume lots of food before going back to bed again. A K-L patient may also run a moderate fever, even though there is no evidence of any current infection, and may exhibit signs of a greatly increased libido such as compulsive masturbation. These attacks generally subside after a few days but may repeat over the course of many years.[1] After recovering, a patient may only vaguely recall what he has been doing, as if he had been in a dreamlike state. Occasionally, these attacks may take place not long after a patient has had the flu, which may thus have a role in triggering this disorder.

The reasons for the bizarre behaviors seen in the Kleine-Levin syndrome are not really known. If you have been paying attention to most of this book, you will recognize that many of the symptoms—overeating, sleepiness, fever, and sexual excitement—could be due to an abnormal function of the hypo-thalamus, which regulates all of these things. Another peculiar feature of the K-L syndrome is that three-fourths of the patients are male and the onset of the disorder typically takes place during the teenage years when the hormonal changes of puberty occur.

Since exposure of the hypothalamus to steroid hormones such as testos-terone or progesterone can provoke sexual excitement, sleepiness, increased appetite, and increases in body temperature, I speculated in my first paper, published when I was a young and inexperienced graduate student, that an abnormal response of the hypothalamus to sex steroids might be a component of this disorder.[32] In many ways, the Kleine-Levin syndrome is a "mirror image" of anorexia nervosa, involving mainly males instead of females, an increase in appetite rather than a decrease, and bouts of sleepiness rather than bouts of compulsive exercise, all taking place at the time of puberty. Perhaps another type of abnormality in the hypothalamic response to sex steroids, similar but opposite to what may take place in anorexia, might explain the syndrome. However, this syndrome is much rarer than anorexia nervosa, so it is much more difficult to study it or find the cause of it.

Some support for my thinking has unexpectedly emerged from a few studies of female patients with the Kleine-Levin syndrome. In these stud-ies, periodic attacks of sleepiness and increased appetite coincided with a particular phase of the menstrual cycle. They could be abolished by treatment with synthetic hormones that stopped ovulation and halted the production of sex steroids by the ovaries.[2,16,21] One of these studies was performed in the lab of Dr. William Dement, who, as noted earlier, played a leading role in the understanding of narcolepsy, so I feel pretty confident in these data in particular.[2] So, perhaps an abnormal hypothalamic response to progesterone (produced by the ovaries) or testosterone (produced by the testicles) may underlie the Kleine-Levin syndrome. It may never be possible to directly test this hypothesis about the Kleine-Levin syndrome on real patients, since the disorder is so rare, but at least the idea seems plausible and gives some cred-ibility to the overall proposal that an abnormal response of the hypothalamus to sex steroids could produce unusual behaviors.

WHAT IS SLEEP GOOD FOR?

Even though we are beginning to understand the neural circuits that provoke sleep, it is still not very clear yet what the benefit of sleep actually is. As a matter of fact, you could propose many reasons why sleep could potentially be harmful to a human or another type of animal: when asleep, an animal is helpless and poorly responsive to threats in his environment (e.g., lions or tigers), he misses opportunities to find food or a mate, and he is generally vulnerable to lots of things. Why sleep at all?

There must be some good reason for sleep, because all animals do it. Brian Preston at the University of Durham in England, along with his co-workers, has come up with a pretty convincing argument that sleep helps animals fight off infections. He did this by looking at animals that sleep for only three or four hours per day (e.g., sheep) and animals that sleep for longer periods (e.g., chimpanzees, which sleep for 10 hours each day, or hedgehogs, which sleep for an astounding 18 hours per day). Animals that don't sleep very much tend to be animals that are constantly fearful of predators and thus must always be on the alert for them (sheep are good examples of this). Preston found that the activity and number of immune cells (e.g., types of white blood cells called neutrophils) in animals that sleep very little were substantially lower than in animals that sleep a lot during the day. He noted that, when we get an infection, the immune system generates molecules that prompt the hypothalamus to initiate a fever and to stimulate sleep more than usual. Finally, when rats are sleep deprived for days, they eventually succumb to bacterial infections. All of these data suggest that when we sleep, the bodily energy that would otherwise be devoted to running around and exploring our environment is instead diverted into strengthening the immune system and protecting us from infections.[18]

Another possible benefit of sleep is that it seems to help the brain con-solidate memories of things we are trying to learn and also to dispose of mem-ories that are not helpful. Quite a few studies have shown that if people are deprived of sleep, they learn specific tasks less well. Also, following sessions in which people are asked to learn a new activity, they experience a greater than normal amount of REM sleep that decreases back to baseline once the activity has been fully learned. So, it seems likely that sleep has helpful effects on the brain as well as the body.[19]

CHEMICALS THAT CAUSE SLEEPINESS—HOW ADENOSINE REGULATES SLEEP AND HOW COFFEE WAKES US UP

One aspect of sleep that has long puzzled scientists is that when a person stays awake for a long time, the urge to sleep becomes greater and greater until that person finally has to give in to it. This suggests that some chemical may gradually accumulate in the awake brain over time until levels of it are so high that sleep occurs. One likely candidate for this mysterious sleep-inducing chemical is a small molecule called adenosine, a compound that is produced in increasing amounts when cells become metabolically active. When the brain is highly active and using lots of energy, concentrations of adenosine steadily increase within the fluids bathing brain cells. In the sleeping brain, metabolic rates are about 30% lower than during an awake state; this allows cells to dispose of adenosine, which therefore plummets during a refreshing sleep.

Enzymes that metabolize adenosine are particularly abundant in the tuberomammillary nucleus of the hypothalamus, the cluster of cells that utilizes the neurotransmitter histamine to keep us awake. It seems likely that the reason adenosine can cause sleepiness is that it suppresses the firing of these alertness-producing tuberomammillary neurons. We do have one convenient weapon to battle these sleep-inducing effects of adenosine, however: the caffeine present in coffee and tea is a potent blocker of adenosine receptors. Caffeine thus can "reawaken" the alertness-producing neurons of the hypothalamus and provide us with the "jolt" we need to stay awake in the early mornings.[12]

Finally, adenosine seems to also play a major role in the creation of sleepiness in hibernating animals. If chemicals that activate adenosine receptors are infused into the brains of arctic ground squirrels during the winter, they cause the squirrels to go to sleep immediately; however, when the same chemicals are infused during the summer, they have no effect! Apparently, seasonal changes in sensitivity to adenosine seem to underlie the dramatic seasonal differences in sleepiness in hibernating animals.[6]

DREAMS AND SLEEP

Dreams are a dramatic component of sleep, but the functions and generation of dreams, amazingly, are still not well understood. What do dreams mean? Can they predict the future, or are they only windows into our past actions? Many scientists have proposed that dreams are a way of rehearsing activities that we have recently learned and will need again, but like many hypotheses about dreams, it is hard to prove that this is true.

People have been fascinated by dreams throughout history. The ancient Greeks believed they were divine gifts from the gods and could serve as guides to the future or to the overall health of the dreamer. One professional dream interpreter named Artemidorus Daldianus compiled a list of 30,000 dreams that he interpreted. Other Greek thinkers like Aristotle belittled these notions and felt that dreams arose from disturbances within the dreamer himself.[14]

Perhaps the most famous modern interpreter of dreams was Sigmund Freud, who published his controversial *Interpretation of Dreams* in 1899. In this massive work, Freud analyzed the hidden meanings of dozens of dreams by himself and others and concluded that many dreams fall into the category of "wish fulfillment," that is, stories that solve the problems we have in everyday life. In one well-known example, Freud described his own dream of riding on a gray horse and meeting one of his medical colleagues on the road. At the end of the dream, Freud dismounted the horse, led it up to his hotel in a town, and then woke up. Freud found this to be a puzzling dream, since he had never ridden a horse in real life. Also, at the time of the dream, he had developed a painful boil on his backside that made even sitting, let alone riding a horse, a painful experience. Freud maintained that the dream illustrated his desire to cure the boil and have done with painful sitting, as well as a desire to talk with his medical colleague, who habitually wore a gray suit, the same color as the horse.

The problem with this type of analysis is that, even though the explanation might seem plausible, many other interpretations are also possible that can't objectively be ruled out by tests or experiments. This puts Freud's analysis of dreams more into the realm of philosophy than neuroscience. Also, Freud tended to impose his views of sexuality upon dream interpretation. For example, when a young woman explained to him that she had dreamed of wearing a strangely shaped straw hat, with its middle piece turned up and its side pieces turned down, Freud responded that he had no doubt that the hat represented the male genital organ. Since this era of dream interpretation, scientists have focused more upon the categories of dream content, the timing of dreams, and the brain structures activated during dreaming.[5]

Certain types of dreams appear to be more common than others. Many of us have experienced some of the most frequent themes in dreams, such as dreaming we are falling from a great height or actually flying, dreaming that we are taking (and failing!) an examination for a class that we forgot to attend, dreaming that we are walking around in public partly or completely undressed, or dreaming that we are having a physical fight or argument with someone. Some of these dreams clearly reflect anxiety about some past event or hypothetical experience.[24] Dreams about flying are puzzling, because they represent an experience that we can never have had (at least not without the aid of an airplane). Some researchers have speculated that dreams of flying may be related to inner ear disturbances. It is well known that flushing the ear canal with warm water can provoke a sensation of vertigo and involuntary eye movements due to heat-induced currents in the fluid within inner ear structures that monitor head position and control overall balance. Perhaps lying on one side on a pillow heats up one ear and creates a conflict with the structures of the other ear, leading to an illusion of vertigo and falling. But even this reasonable interpretation of the content of these kinds of dreams still falls within the realm of speculation.

Dreams are rich in visual and auditory sensations but rather poor in smells and tastes. In accord with this, dreams seem to be generated by activity in the frontal and parietal lobes of the brain, which are important for the analysis of visual and auditory stimuli. If input to these regions of the cortex is cut, as occurred in patients who underwent a procedure called prefrontal lobotomy that was a common practice in the 1950s for the treatment of psychosis, the ability to dream, as well as the ability to fantasize and plan future actions, is dramatically reduced.

Curiously, although infants and young children sleep much more than adults, if a young child is awakened during REM sleep, he will report a dream only about 20% of the time. Adults, in contrast, report dreaming 80% of the time when awakened during REM sleep. Moreover, the dreams of young children are less complex than those of adults and rarely involve the child as an active director of a dream sequence. Differences between adult and childhood dreaming may be related to the slow process of maturation of the cortex over the years. For example, nerve fibers that project to the parietal cortex are not completely insulated (myelinated) until about the age of seven, so this brain region takes a while to develop the ability to generate complex dreams.[10]

When we are dreaming, it is very important that we not act upon our dreams and become injured, so the activity of motor neurons within the spinal cord that innervate our muscles is suppressed. It has taken quite a long time to identify how this sleep-related paralysis, or atonia, comes about. The most recent research suggests that there are specific populations of neurons in the pons and medulla that send fibers "downstream" in the brain to depress the activity of spinal motor neurons. Some of these neurons are regulated by orexin, the neurotransmitter produced in hypothalamic cells that also regulate arousal and sleep.[7,29]

Occasionally, injury to the pons, due to an accident or a stroke, can damage these sleep- and paralysis-regulating neurons. This condition is called REM sleep behavior disorder. One patient with this medical condition experienced terrible trials when he went to sleep. When he dreamed, he would scream and thrash his arms and sometimes would fall out of bed or punch and kick his wife. One time he dreamed that he was being chased by a wolf; his sleep was interrupted by screams from his wife, whom he had punched in the nose, thinking he was hitting the wolf with a stick.[30] Fortunately, this condition is relatively rare.

One possible function of sleep and dreaming, as we have said, is aiding the consolidation and formation of memories. One of the surprising recent findings about memory formation is that this process seems to require the formation of brand-new neurons in a part of the brain called the hippocampus. In this brain region, thousands of new neurons are formed from precursor "stem" cells every week, and many of these promptly become incorporated into the circuitry of the hippocampus. In almost all of the rest of the brain, new neurons are never formed, so that if the neurons of the cortex that we are born with get damaged, the cortex cannot repair itself, and the resulting neurological deficits become permanent. So, the ability of the hippocampus to form new neurons, and the involvement of these new neurons in memory formation, has attracted a lot of attention recently. If sleep and dreaming are experimentally prevented, memory formation and the formation of new neurons in the hippocampus are seriously impaired.[14]

All of these stories show that, although we spend about one-third of our lives asleep, the reasons for sleep and dreams and the hypothalamic circuitry that controls sleep have only begun to be understood relatively recently. We are only at the threshold of comprehending the functions and meaning of sleep.

· 5 ·

The Hypothalamus and
the Control of Hormones

$\mathscr{I}$t must seem obvious to you readers that the hypothalamus plays some role in controlling the secretion of hormones because it is so close to a major endocrine organ, the pituitary gland (also called the hypophysis), which hangs beneath the hypothalamus from a piece of tissue called the median eminence. While this is certainly clear today, it was far from obvious to previous generations. The pituitary gland is buried so deeply beneath the brain that early scientists had little access to it and could only speculate about what exactly it might do. Some thought that it merely removed mucus from the nearby nasal cavity.[27] The name of the pituitary gland itself is derived from the Latin word for phlegm (*pituita*).[20, 21]

The first real indication that the pituitary might be good for something more than regulating phlegm arose from the studies by a French neurologist named Pierre Marie, who was interested in a strange disease called acromegaly. In acromegaly, a person gradually develops enlarged hands and feet, a curvature of the spine, and increases in the size of his nose and ears that slowly make his facial features coarser and less attractive. These disfiguring and painful symptoms can sometimes be accompanied by changes in vision. In 1886, Marie noted that these peculiar features were often accompanied by an enlargement of the pituitary gland, and he suggested that the pituitary might contain some growth-promoting substance (we now call this substance growth hormone). Too much growth hormone causes an overgrowth of the soft tissues of the adult body seen in acromegaly.[16] Also, an enlarged pituitary gland can press upon the nearby optic nerves and gradually cause blindness.

For years, scientists were baffled by the question of how to test this proposal that the pituitary might contain a growth-promoting substance. One logical approach would be to remove the pituitary and see if growth of an animal is affected. This, however, is easier said than done. How could the pituitary be removed if it lies *beneath* the brain?

A Romanian physiologist named Nicholas Paulesco was one of the first to try to overcome this daunting challenge by opening up the cranium of an anesthetized dog, gently lifting up one lobe of the brain to expose the pituitary, and pulling out the gland before allowing the brain to settle back into the skull again. Needless to say, many of the dogs quickly died after such a drastic surgery, and Paulesco could only conclude that the pituitary was necessary for life.

Other scientists such as Harvey Cushing at Johns Hopkins in Baltimore created different surgical approaches to the pituitary that were less dangerous. One approach, which Cushing applied to over 200 patients with acromegaly, is to make an initial incision below the brain rather than above it.

In the so-called transsphenoidal approach to the pituitary, the surgeon starts out by making an incision into the mucous membrane just above the front upper teeth. Then this mucous membrane is slowly pulled upward and away from the underlying bone until the mucous membrane and skin of the upper lip become pushed up onto the tip of the nose. This allows access into the nasal cavity just above the mouth. Probes and scalpels are then pushed into the back of the nasal cavity, the sphenoid bone that encloses the pituitary is cut open, and overgrown parts of the pituitary can be removed. This was shown to stop the overgrowth of tissues in acromegalic patients.[16] Also, because the outer surface of the face needn't be incised, this type of surgery avoided disfiguring scarring.

This approach worked fine for humans, but it wasn't very useful when it came to small experimental animals like rats. Other surgical approaches were gradually developed that allowed scientists to discover more and more pituitary hormones that affect many different organs in the body. Thus, the pituitary changed from an object of disinterest into a "master gland" that secreted unexpected quantities of hormones. The bewilderment and excitement of this time was summed up nicely by Phil Smith, an anatomist working at Columbia University in New York who had developed his own method of hypophysectomy for rats. In 1935 he noted that "it is evident that no less than six and possibly eight hormones have been extracted from the anterior pituitary. That this small gland, which in man averages less than 0.5 grams in weight, secretes this number of hormones . . . taxes the imagination."[27]

The use of rats and other animals in endocrinology experiments allowed scientists to confirm that many of the symptoms seen in human patients with damaged pituitaries (decreased growth as well as shrinkage of the adrenal glands, ovaries, testes, and thyroid glands) must be due to a removal of the hormones that are normally produced by the pituitary gland. What hormones, then, are found in the pituitary, and how does the hypothalamus control the activity of this important endocrine organ?

HORMONES OF THE PITUITARY

The pituitary is actually two glands that have fused together to form a single organ. Each of these two portions or lobes of the pituitary produces quite different hormones.

The anterior lobe of the pituitary gland begins its life during embryogenesis, when a pocket of tissue called Rathke's pouch grows upward toward the brain from the developing nasal cavity. This pouch of tissue will become the anterior lobe of the pituitary. At the same time, a portion of the developing hypothalamus sends a mass of tissue called the infundibular stalk down toward the mouth; this stalk will form the posterior lobe of the pituitary. Occasionally this developmental process will go astray and portions of Rathke's pouch will fail to fuse with the posterior lobe of the pituitary. These lost portions can sometimes become cancerous, forming tumors called craniopharyngiomas that can enlarge and damage the functional parts of the pituitary.

If you take a look at the anterior lobe of the pituitary, you can distinguish two types of cells that stain quite differently with the right sorts of stains. One cell type is called an acidophil cell because its cytoplasm attracts acidic biological stains and stains light pink. Basophil cells, in contrast, tend to stain dark pink or purple with more basic stains. These two cell types stain differently because the cytoplasms of these cells are chock full of vesicles that contain hormones. Hormones of the acidophil cells attract acid stains, and hormones of the basophil cells attract basic stains because these protein hormones become decorated with very acidic chains of carbohydrate molecules. These chemical differences between the hormones make the cells that produce them stain differently.

HORMONES PRODUCED BY PITUITARY ACIDOPHIL CELLS

Which hormones do these two cells make? The acidophil cells make one of two hormones: prolactin, which stimulates mammary glands to synthesize milk, and growth hormone (GH), which stimulates the enlargement of both bones and soft tissues.

Let's stop here for a minute. The last sentence above is actually not true: growth hormone does *not* cause growth directly, in spite of its name! This surprising fact was first discovered in 1957 by Dr. William Daughaday, who was working in a lab at Washington University in St. Louis. At the time, he was trying to measure the growth-promoting effects of crude growth hormone preparations extracted from the pituitaries of cows. To do this, he measured the growth and sulfate uptake of pieces of cartilage incubated in a dish. To his surprise, when GH was added to the cartilage cultures, nothing happened!

If it had been up to me, I would have concluded that I had made a mistake, and I would have thrown out the experiment. But Daughaday didn't jump to this conclusion. Instead, he added some blood plasma into his cultures and found that growth was stimulated by 200%! However, plasma from hypophysectomized rats without pituitaries completely failed to stimulate growth. What in the world was going on?

Daughaday eventually concluded that growth hormone causes growth *indirectly*, by promoting the synthesis of a secondary factor now called insulin-like growth factor 1 from tissues like cartilage and liver that was then carried in the bloodstream. This explained the lack of growth-promoting effects of blood obtained from rats that lacked a pituitary and couldn't make growth hormone.[10]

The other hormone of acidophil cells, prolactin, does have direct effects upon mammary gland epithelial cells and does indeed directly stimulate milk synthesis by binding to prolactin receptors on these cells. It turns out that these two acidophil cell hormones have amino acid sequences that are rather similar to each other. This suggests that, way back in time, there was a single gene for these hormones that coded for a single protein. At some point, this gene was duplicated and gradually acquired mutations in it that led to changes in the amino acid sequences of the proteins they coded for. Thus, two rather than one hormone became the products of acidophil cells.

HORMONES PRODUCED BY PITUITARY BASOPHIL CELLS

The basophil cells make a wider variety of hormones. Three of these hormones (TSH, LH, and FSH) are also very similar in structure to each other; these separate hormones also probably arose when the gene for one common ancestral protein was copied into three chromosomal copies instead of one and then acquired mutations that changed the functions of the three proteins coded for by the new genes.

TSH (thyroid stimulating hormone) stimulates the thyroid gland to make its hormone, thyroxine. LH and FSH are so-called gonadotropins and specifically affect the functions of the gonads (ovaries or testes). FSH acquired its name (follicle stimulating hormone) because it causes the enlargement of ovarian follicles, cellular structures that encircle egg cells within the ovary. Follicle cells are the source of the steroid hormone estrogen. LH acquired its name (luteinizing hormone) because it stimulates the rupture of ovarian follicles during ovulation, which releases egg cells from the ovary into the fluid in the abdominal cavity. After a follicle ovulates, the cells that remain behind in the ovary become transformed into a structure called a corpus luteum. These new, luteinized collections of cells make progesterone, the steroid hormone that maintains the uterus in a state receptive for a possible pregnancy. Each hormone is made by one variety of basophil cell.

A fourth variety of basophil cell (also termed a corticotroph cell) makes entirely different hormones from an entirely different precursor molecule. We have already encountered this protein, called proopiomelanocortin (POMC), before: it is used as a neurotransmitter by feeding-restraining neurons of the arcuate nucleus of the hypothalamus (see chapter 1). In the pituitary, this same protein is produced by basophils and secreted into the bloodstream as a hormone. Before being secreted, this protein is chopped into three fragments. One fragment is called ACTH, for adrenocorticotropic hormone. This protein hormone is carried by the bloodstream to the adrenal glands, where it stimulates cells of the adrenal cortex to make steroid hormones. A second fragment is called MSH, for melanocyte-stimulating hormone. MSH stimulates pigment-containing cells of the skin called melanocytes to synthesize melanin and add a brown tone to the color of the skin. A final fragment, called β-endorphin, binds to opiate receptors on pain-regulating neurons of the nervous system. This hormone has the ability to reduce sensitivity to pain, just like morphine or opium can. The hormones of the basophil cells are released during painful or stressful conditions that require an activation of the adrenal gland and a lessened sensitivity to pain.

One noteworthy feature of almost all of the hormones secreted by the anterior pituitary is that they really have no effect on most of the cells in the body. They only serve to activate other endocrine organs that do regulate the function of muscle, fat, and other tissues. So, for example, LH only activates the ovaries, which respond by secreting estrogen, and TSH activates the thyroid, which responds by secreting thyroxine. Thyroxine and estrogen not only alter the function of muscles and fat cells, but they also do one other important thing. They exhibit an action called *negative feedback* by going back up to the pituitary and hypothalamus and dampening the activities of these organs. So, if the hypothalamus orders the pituitary to secrete too much LH, a rise in blood levels of estrogen feeds back to the hypothalamus and pituitary and prevents more LH from being secreted. This helps maintain levels of reproductive hormones within reasonable and physiologically appropriate bounds.

The principle of negative feedback even applies to the overall growth of the body: if too much growth hormone is secreted, it causes an inappropriately high blood level of insulin-like growth hormone, which feeds back upon the hypothalamus and pituitary to suppress further secretion of growth hormone. This very important property of anterior pituitary hormones allows for a tight control of the endocrine system.

The posterior pituitary is not really a gland at all but a mass of nerve endings that are formed by axonal processes that originate in the supraoptic and paraventricular nuclei of the hypothalamus (Fig. 2-1). When the nerve cells in these nuclei are activated, they prompt the release into the blood of either vasopressin or oxytocin from the nerve endings of the posterior lobe of the pituitary. Other nerve cells in these nuclei, as we have seen, utilize these hormones as neurotransmitters and send them to other portions of the brain to regulate feeding behavior and emotional behavior (see chapters 1 and 3).

It is thus pretty easy to understand what controls the release of hormones from the posterior pituitary. But what about the anterior pituitary? Many of its hormones are secreted in a steady and stable fashion, but all of them show variations in blood levels depending on the hour of the day, and some of them (LH, for example) can be secreted in a massive burst from the pituitary in special circumstances (a surge in blood levels of LH that occurs once a month triggers ovulation in women). What controls the release of anterior pituitary hormones?

THE HYPOTHALAMIC CONTROL
OF THE ANTERIOR PITUITARY

An early clue about these things was provided by Dr. Percival Bailey, a staff member working for Harvey Cushing in Baltimore. In 1920, Dr. Bailey was attempting to perform a hypophysectomy on an anesthetized dog by going at the pituitary through the roof of the dog's mouth, but he accidently cut a small artery that caused a lot of bleeding. He gave up on the procedure and sewed the dog up for recovery. The next morning, he was surprised to see that the dog had urinated profusely onto the floor of its cage, and the dog continued to do this for weeks. He also detected abnormalities in body weight and hormone secretion in the dog. Since he had not touched the pituitary itself and had only punctured the hypothalamus, he concluded that the hypothalamus must have some controlling influence on the pituitary and published a paper describing his ideas. Dr. Cushing, who thought the pituitary functioned independently of the hypothalamus, was furious and for a while tried to stop the publication of the paper! Later on, he relented and realized his ideas were wrong.[1]

If the hypothalamus influences the pituitary, how exactly does it do this? One possibility was by the release of unknown chemicals into blood vessels that connect the hypothalamus to the pituitary (pituitary portal vessels). This idea, however, was ridiculed by people who thought these vessels merely deliver blood to the brain from the pituitary. Two English physiologists at Cambridge, John Green and Geoffrey Harris, decided to resolve this dispute once and for all by looking at the direction of blood flowing in these tiny vessels in a live, anesthetized rat. Once they accomplished this difficult task, they reported that blood mainly flowed from the hypothalamus to the pituitary, not the other way around. Later, they transplanted a pituitary into the empty sphenoid bone of a hypophysectomized rat and found that this donor pituitary became rejuvenated and started to secrete its hormones as usual. This didn't happen if they transplanted a pituitary into a pocket beneath the temporal lobe of the brain, so they concluded that some mysterious chemicals must be emanating from the hypothalamus to release hormones from the pituitary.[1]

The ensuing race to discover the hypothalamic releasing factors that control the pituitary is one of the most dramatic stories in endocrinology. It basically began in the lab of Dr. Roger Guillemin in Houston. Dr. Guillemin, who had emigrated from France to the United States, realized, from all the evidence then available, that the amounts of the postulated releasing factors contained within the hypothalamus must be very small. Accordingly, he began collecting huge numbers of hypothalamic chunks from cow heads acquired at slaughterhouses. This is not an easy task: to obtain each hypothalamus, the thick skull of a cow must be cut open with an ax, the brain must be removed, and then the tiny fragment of hypothalamus must be dissected free of other tissues, cleaned, and frozen. Between 1964 and 1967, about 5 million chunks of hypothalamus, amounting to almost 50 tons of tissue, were collected by Guillemin and his collaborator, Wylie Vale.[17]

To analyze these tissue fragments, about 10,000 pieces of hypothalamus at a time were frozen, ground into powder, and chemically treated in attempts to extract the releasing factors. One scientist who helped in this effort was Andrew Schally, who like Guillemin had been born in Europe (Schally's homeland was Poland). For four years, Guillemin and Schally attempted to isolate the factor that stimulated the release of ACTH from the pituitary. Despite many promising beginnings, they were at first unsuccessful, and Schally left the lab to start up his own research enterprise in New Orleans at Tulane University. Schally got the help of the Oscar Mayer company, which donated 1 million fragments of pig hypothalami to his research effort. Both scientists continued the quest for the factors, now independently and in competition with each other.

Finally, in 1962, the first hypothalamic releasing factor was identified. It was called TRF, for the factor that releases TSH from the anterior pituitary. Guillemin triumphantly sent his landmark research report to the prestigious journal *Science*, only to have it rejected by a reviewer! The reviewer stated that "hypothalamic releasing factors were not much else than a lasting fancy of Guillemin's vivid imagination." The sting of this unfounded rejection must really have annoyed Guillemin, since he recalled it vividly in a historical paper he wrote 13 years later.[17]

As more and more tissue samples were analyzed, more and more releasing factors were discovered. In 1971, Andrew Schally's team beat the Guillemin team to the punch by discovering the releasing factor for luteinizing hormone, which he called LH-RH. Gradually, releasing factors (now called releasing hormones) for LH, ACTH, GH, and TSH were identified in both labs. Many of these small peptides have not only been localized to specific nerve cells in the hypothalamus but have also been found to be utilized as neurotransmitters in the spinal cord and other regions of the central nervous system. Schally and Guillemin shared the Nobel Prize in 1977 for their heroic work and discoveries. Many of their studies were still under way when I became a graduate student in 1972, so it was an exciting time for those of us interested in the hypothalamus.

Most of the releasing hormones discovered by Guillemin and Schally were small peptides that contained only 3 to 12 amino acids each. Once their structure was known, it became possible to prepare antibodies to them that could be used to stain hypothalamic sections. These studies showed that TRH and CRH are produced by neurons in the paraventricular nucleus. Neurons in this nucleus are extremely sensitive to stressful situations, and when excited by stress, they cause the pituitary to secrete ACTH and TSH. Other paraventricular neurons project way down to the spinal cord to activate the sympathetic nervous system and the adrenal medulla during stress, causing a host of physiological responses (e.g., increased blood pressure and heart rate, increases in blood glucose, and other activities that allow us to respond to a dangerous situation and run away to escape it). Other releasing hormones like GH-IH (growth hormone inhibitory hormone) and GH-RH are found in the arcuate nucleus. We have already seen that a neurotransmitter called dopamine, which diminishes prolactin secretion from the pituitary, is produced by neurons of the arcuate nucleus (see Fig. 1-5). LH-RH containing neurons are present both in the anterior hypothalamus and in the arcuate nucleus.

The story of hypothalamic releasing factors is still not done. Between 2000 and 2006, Lance Kriegsfeld and coworkers at UC Berkeley found yet another small peptide that inhibits the release of LH from the pituitary. The nerve cells that make this peptide are found mainly in the dorsomedial hypothalamus, just above the ventromedial nucleus, and appear to synapse onto neurons that make LH-RH and inhibit their function.[22]

In contrast to these inhibitory cells, another subset of hypothalamic neurons was also recently discovered to increase the function of the reproductive system by stimulating LH-RH containing neurons. These stimulatory cells make a peptide that is structurally similar to the LH-inhibiting peptide and that, appropriately enough for topics in reproduction, is called kisspeptin. Kisspeptin was originally identified in studies of cancer cells and seems to have gotten its name because it was found by Danny Welch and coworkers at Penn State University in Hershey, Pennsylvania (home of chocolate kisses).[23, 25] It seems that scientists can be as whimsical as anybody else when choosing a name for something! If kisspeptin-containing neurons are prevented from making kisspeptin, a rat will not go through puberty and won't be able to reproduce, so it is clear that these cells have a commanding role in regulating reproduction.[41]

All of these discoveries made it much easier to understand how the secretion of hormones from the pituitary is controlled. Still, some aspects of endocrinology resisted an easy explanation until quite recently. One of these was the regulation of LH secretion.

This reproductive hormone is not released steadily and boringly from the pituitary but shows wild gyrations in blood levels over the course of the monthly ovarian cycle of women. In the first half (12 days or so) of the cycle, blood levels of LH slowly increase until they suddenly show a drastic rise called the LH surge. This dramatic rise in LH causes ovulation to occur in the ovaries. Then blood levels of LH fall and remain flat until menstruation and the start of a new ovarian cycle. For many years, I gave lectures about the female reproductive system to my students and would show them a diagram of LH levels over the course of the menstrual cycle.[39] They dutifully copied it down, and I was lucky they did not ask what actually caused these dramatic changes in hormone levels. If they had, I would have had to admit that I didn't have a clue!

To clarify, it has been known for a long time that estrogen, secreted from the ovaries, can influence LH levels. However, it does this in a confusing way: during the first half of the ovarian cycle, low blood levels of estrogen *suppress* LH secretion, but toward the middle of the ovarian cycle, higher blood levels of estrogen *stimulate* LH secretion and cause the all-important LH surge! How could the same hormone have diametrically opposite effects on LH secretion during different days of the cycle?

A recent study, conducted in Kevin Catt's lab at the National Institutes of Health in Bethesda, Maryland, may finally have solved this puzzle. It turns out that there are two types of estrogen receptors in the brain, the alpha receptors and the beta receptors. The alpha receptors respond to low levels of estrogen and suppress the activity of the LH-RH containing neurons that possess this type of receptor. The beta receptors are found in another population of LH-RH containing neurons and respond to higher levels of estrogen by stimulating the activity of their LH-RH containing neurons.[20]

After ovulation and the release of an egg cell from an ovarian follicle, the corpus luteum of the ovary secretes progesterone, which suppresses LH once again and prevents a second LH surge from occurring. A corpus luteum only lasts for about 14 days and then degenerates, causing a fall in blood levels of progesterone, the onset of bleeding in the lining of the uterus, and the beginning of a new ovarian cycle. The revelation that there are *two* types of LH-RH containing neurons that respond to estrogen in opposite directions finally allows me to make an intelligent explanation to my students of the changes in LH seen during the ovarian cycle, and more importantly, it allows scientists to better understand how to control ovulation and fertility.

Another puzzling question about the reproductive system was that, if humans or animals experienced a famine and lost body weight, they would show declines in reproductive ability, and females would stop ovulating. But what was the mechanism for this loss of fertility during a famine? We now know that leptin, secreted from fat cells, normally excites kisspeptin-containing neurons in the arcuate nucleus. These activated kisspeptin-containing cells, in turn, promote the release of LH and enable ovulation to occur. If fat cells become shrunken during a famine and stop producing leptin, this whole excitatory circuit in the hypothalamus stops working, and reproduction ceases.[3,32,45]

THE HYPOTHALAMUS AND GLUCOSE-REGULATING HORMONES

Control of the pituitary by the hypothalamus is important, but it is by no means the only contribution of the hypothalamus to endocrinology. Neurons of the hypothalamus also regulate other hormones that affect levels of blood sugar (glucose). What are these hormones?

One important glucose-regulating hormone is adrenalin (epinephrine), a small molecule secreted from the inner portion (medulla) of the adrenal gland, an endocrine organ perched on top of the kidney deep within the abdominal cavity. We all know what it feels like to experience an "adrenalin rush": during frightening or stressful situations, adrenalin is poured into the bloodstream and causes a heightened sense of consciousness, an increased heart rate, and often a temporary surge in strength. I can vividly recall the last time I myself felt such a "rush," several winters ago when my car slid on "black ice" and I suddenly feared that I would collide with oncoming traffic! I managed to regain control of the car but felt jittery and breathless for some time afterward.

These effects of adrenalin help us to cope with competitors during a road race or with falling off of a cliff or being chased by a bear. Other effects of adrenalin, however, are also important. When blood glucose falls dangerously low, adrenalin is released into the blood to stimulate an increase in blood sugar, which is poured into the circulation from the liver in response to adrenalin. This protective reflex release of adrenalin during hypoglycemia is particularly important for the brain, which is virtually unique among our organs in that it is almost totally dependent upon glucose as a fuel source. Without enough glucose, the brain stops functioning well, and we pass out or even die. This can indeed happen if an episode of hypoglycemia is severe enough.

What prompts this protective release of adrenalin from the adrenal gland during hypoglycemia? The signal does not arise within the adrenal gland itself, because the cells of the adrenal gland lack a glucose-sensing mechanism. What do I mean by that?

It turns out that several organs in the body continuously monitor blood concentrations of glucose and react vigorously when levels of this critical nutrient rise or fall. One of these organs is the liver. The liver receives nutrient-rich blood directly from the vasculature of the intestines. When we drink a chocolate shake at McDonald's, the sugars in the shake are digested and funneled directly to the liver.

Liver cells possess specialized glucose transporter proteins on their cell membranes called GLUT2 transporters (there are about 13 varieties of GLUT transporter proteins, all with different properties and different cellular locations). These GLUT2 transporters only become active when blood sugar levels become very elevated. These transporters push glucose into the liver cells, where it becomes stitched together into long chains of sugars that are called glycogen molecules. This allows the liver to accumulate a storage form of glucose. Later, whenever blood levels of glucose become too low (hypoglycemia), the glycogen in the liver can be broken down again into glucose, which is then released back into the bloodstream. Adrenalin protects the body from hypoglycemia by stimulating liver cells to release their glucose.

It is really a good idea for liver cells to use GLUT2-type glucose transporters, because if they used the type found in nerve cells, for example GLUT1, the liver would continuously suck glucose out of the bloodstream even if blood glucose levels were relatively low. As we have seen, this would kill the brain.

However, let's get back to the basic question of how cells of the adrenal medulla "know" that blood glucose has fallen too low. These cells do *not* possess GLUT2 transporters and cannot by themselves measure levels of blood glucose, so how do they "know" how to react when blood sugar levels fall too low? It turns out that the adrenal cells are activated by nerves of the sympathetic nervous system to swing into action during hypoglycemia. But from where do these nerves get their commands?

One way to find out is to subject a rat to mild hypoglycemia by giving it a small shot of insulin, a hormone that lowers blood glucose (more on this later!), and then overdosing the rat with anesthetic so that the brain can be removed. Subsequently, the brain can be examined for cells that have a special sensitivity to glucose and have become injured by hypoglycemia. Barry Levine and coworkers at the VA Medical Center in East Orange, New Jersey, performed this study some years ago and found that neurons specially sensitive to hypoglycemia were located in only two places: in a small area in the medulla of the brain and in the arcuate nucleus of the hypothalamus.[40] During hypoglycemia, these neurons activate circuits, routed through the paraventricular nucleus, that control the adrenal gland and stimulate the protective release of adrenalin. Levine also proposed that recurrent bouts of hypoglycemia, like that experienced by patients with poorly controlled diabetes mellitus, may cause permanent damage to these glucose-sensitive neurons and may thus lead to a vicious circle, in which an impaired reaction to hypoglycemia causes neuronal damage that leads to additional, worse responses to hypoglycemia.

The proposal that these hypothalamic neurons are specially sensitive to glucose can also be confirmed by another approach, namely by creating mild hypoglycemia in a rat and seeing which neurons in the brain show hyperactivity in response and thus stain for the c-fos protein. So, it is clear that glucose-responsive neurons in the arcuate nucleus of the hypothalamus have a critical role in preventing episodes of hypoglycemia by stimulating the release of adrenalin.

Adrenalin is not the only important glucose-regulating hormone. Hypoglycemia also stimulates the release of another hormone called glucagon from endocrine cells in the pancreas. What is glucagon, and where, precisely, is it made?

Glucagon is a protein hormone produced by cells called alpha cells that are located in peculiar ball-shaped masses of cells that account for only about 5% of the volume of the pancreas. Most of the cells in the pancreas are called acinar cells. These cells are arranged into irregular clusters called acini that pour fluid into myriads of tiny ducts that ultimately deliver their contents into the interior of the small intestine. These cells do *not* secrete hormones and are devoted to an entirely different job: they prepare digestive enzymes that, for example, chop up ingested proteins into amino acids. This fact was known for many years. It was never dreamed that the pancreas could also produce hormones until a medical student named Paul Langerhans looked closely at sections of the pancreas under the microscope.

Paul Langerhans was a 22-year-old student in Berlin when he was told by his superiors in 1869 that he must complete a research project to graduate. He responded by looking at sections of a rabbit pancreas and noticed that he could occasionally find small, pale-staining balls of cells in the pancreas that attracted an unusually rich mass of nerve endings. Langerhans died in 1888 without ever knowing the function of his newly discovered cells. It was not until the turn of the century that it was recognized that the islets of Langerhans, named in his honor, actually were accumulations of hormone-secreting cells.[27]

The islet cells that secrete glucagon (alpha cells) become activated during hypoglycemia. This is appropriate, because glucagon is another hormone that commands the liver to release glucose into the bloodstream. Glucagon also diminishes the uptake of glucose by other cells such as muscle cells. So, glucagon and adrenalin both contribute to the emergency response to hypoglycemia and raise blood levels of glucose back up to normal.

The puzzling thing about the alpha cells is that they, too, lack GLUT2 transporters and thus do not seem to possess the basic elements of a glucose-sensing mechanism. So, how are they activated by hypoglycemia? Bernard Thorens, a research scientist in Switzerland, determined that these alpha cells must also be activated by glucose-responsive cells in the hypothalamus.[7] He found that if GLUT2 proteins were deleted from cells within the nervous system, hypoglycemia no longer provoked glucagon secretion.

One surprising result of Dr. Thorens' experiments was that the glucose-sensing cells of the nervous system did not appear to be neurons at all: if GLUT2 transporters were deleted from neurons, the brain responded to hypoglycemia just fine! However, if GLUT2 transporters were deleted from astrocytes, the brain no longer noticed that blood glucose levels had fallen. It now seems that hypothalamic astrocytes are important monitors of blood glucose.[28] When blood glucose concentrations fall too low, these astrocytes become excited and produce molecules that cause adjacent neurons to swing into action.[26,48] I had some idea that astrocytes might be important for all this about 10 years ago when I performed an experiment in which I gave a drug to rats that injures astrocytes, but not neurons, and found that the regulation of blood glucose by the brain became impaired.[46] Thorens' data, however, are much more conclusive than mine were.[26]

A role for astrocytes in glucose regulation by the hypothalamus is just another example of how specialized hypothalamic astrocytes function as chemical sensors. Hypothalamic astrocytes also respond to changes in blood sodium to provoke thirst (see chapter 2).

INSULIN, DIABETES MELLITUS, AND THE HYPOTHALAMUS

Glucagon is not the only islet hormone that is regulated by the hypothalamus. Other cells in the islets, called beta cells, make another hormone called insulin. The basic task of insulin is to stimulate the uptake of glucose by muscle, liver, and fat cells. This, naturally, causes a fall in blood levels of glucose, particularly after a meal. Thus, glucagon and insulin, by exerting opposing effects on blood glucose, make a combined effort to precisely regulate blood levels of this important nutrient.

Beta cells that make insulin *do* have the ability to respond directly to changes in blood glucose, because they, like liver cells, also possess GLUT2-type glucose transporter proteins. So, beta cells can function as independent glucose-sensing cells, and they generally only secrete insulin when blood levels of glucose are very high. This is a good thing, too, because if they secreted lots of insulin all the time, levels of blood glucose would fall into the basement, we would experience hypoglycemia, and our brains would die. All of this does not mean, however, that secretion of insulin by beta cells is simply a response to blood levels of glucose. These cells, it turns out, are also regulated by the nerves and by the hypothalamus. In fact, an abnormal regulation of islet cell function by nerve endings may be an important component of a serious disease called diabetes mellitus.

There are two types of diabetes mellitus. In the first type, called juvenile-onset diabetes (because it begins at young ages), the beta cells of the islets come under attack by cells of the immune system and eventually die. This drastically diminishes the ability of the pancreas to secrete insulin. As a result, blood levels of glucose rise to very high levels. These levels are so high that absorptive cells in the kidney just can't recover the glucose from the filtrate of blood that we call urine. So, when urine is produced by the kidney in diabetics, it contains very high levels of glucose and, also, very high amounts of water. If untreated, a person with type I diabetes will excrete so much glucose in the urine that all the tissues of the body will degenerate, he will lose drastic amounts of weight, and he will die. To preserve the life of a person with this type of diabetes, insulin must be taken regularly throughout the day. About 2 million people in the United States alone have this disease. While insulin therapy can save their lives, it is tricky to administer just the right amount of insulin each day. Too little insulin fails to relieve the symptoms of diabetes, and too much can cause episodes of hypoglycemia that can potentially damage the brain or even cause death.[14]

I have had experience treating an experimental type of type I diabetes in rats when I was a graduate student. To begin my experiment, I first had to make rats diabetic. It is actually rather easy to make a rat develop type I diabetes. All you need to do is insert a hypodermic needle into the abdominal cavity of the rat and inject a chemical called alloxan. This small molecule has a shape that resembles the shape of a glucose molecule. Because of this, beta cells of the pancreas are tricked into absorbing it via their GLUT2-type glucose transporter proteins. Once inside the cell, however, alloxan cannot be metabolized like molecules of glucose to provide energy. Instead, it has toxic effects that kill the beta cells. Other cells in the body don't take up alloxan very much and thus are not harmed by it.

I vividly recall what happened in the days after I gave rats injections of alloxan. The rats looked normal, but fairly rapidly they became thirstier than normal and started to drink more out of their water bottles, an event that I dutifully measured as part of my PhD thesis. A normal rat will drink only about 30 mL of water a day, but within a week after an alloxan injection, my rats were drinking about 100 mL of water per day and soon began drinking even more than that! The metal pans beneath their cages, which normally contain only dry sawdust to collect their droppings, soon began to overflow with urine, which made taking care of them much more difficult than normal for me. Because this urine contained a lot of glucose, these rats were also losing lots of calories from their bodies and soon began to eat twice as much food as normal.

I was determined to also record their food intake to document this, but if a rat was a messy eater, he would tend to spill some of his rat chow through the wire mesh floor of his cage. Normally this wasn't a problem for me because I could put a paper towel underneath his food dish, collect the spillage and weigh it, and thus have an accurate measurement of how much the rat ate that day. For diabetic rats, however, the paper towels became wet with urine, and I had much more trouble retrieving them, drying out the residue, and weighing the spilled food. I guess this was the penalty I had to pay when I made the rats diabetic in the first place.

My ultimate task as a graduate student was to cure the rats of this condition by giving them injections of insulin and seeing how many doses it took to return them to normal. After giving insulin injections for a week, their water intake returned to normal, their urine output was much reduced (a great benefit both for the rat and for me!), and their daily food intake also began to get lower. It was amazing how powerful the effects of only small doses of insulin were. More than anything else, this illustrated to me the potency of hormones in controlling the body.

So, it is pretty easy to produce and understand type I diabetes in a rat. But what causes it in humans? Why would the immune system destroy cells that are so crucial for the function of the body? All the answers to this question are not known, but a type of mouse that develops type I diabetes may be providing some of them.

The type of mouse I am talking about is the so-called nonobese diabetic (NOD) strain of mouse, which was developed by inbreeding mice in the lab of a Japanese scientist named Makino in 1992. These mice have all sorts of problems with their immune systems and regularly develop a type of diabetes that is very similar to human type I diabetes. A team of investigators headed by Michael Dosch in Toronto, Canada, has recently discovered that sensory nerve endings that innervate the islets of NOD mice have a critical role in the development of diabetes. They found that these nerve endings aggravate inflammation and autoimmune attacks in the islets of these mice and that these attacks could be prevented by injecting a neurotransmitter called substance P into the arteries of the pancreases of these mice.[35] Perhaps a similar adjustment of the function of pancreatic nerves in humans might someday lead to a cure for this type of diabetes.

The other type of diabetes mellitus, called type II or adult-onset diabetes, is much more common than type I diabetes and has a fundamentally different character. In the early stages of type II diabetes, the beta cells of the pancreas are completely normal and produce insulin normally. The problem in type II diabetes is that tissues such as muscle and fat become resistant to the effects of insulin. To overcome this resistance, the beta cells must gradually produce more and more insulin so that finally they become exhausted and damaged.

Type II diabetes mellitus is a serious and widespread disease. About 5% of all adults in the United States are now diagnosed with this disease. Type II diabetes is associated with a doubling of overall mortality rates from a variety of causes and is a leading cause of cardiovascular and kidney disease. The reason for this is that capillaries don't function well in diabetics. This causes functional problems for the capillaries in the kidney that filter blood to produce urine and also for capillaries in places far from the heart like the feet and toes. Blood circulation in the feet is poorer than in most regions of the body, so feet don't receive as abundant a blood supply as most parts. In diabetics, capillary problems in the feet can lead to the death of tissues in the toes, causing bacterial infections, the development of gangrene (blackened and dying toes), and sometimes an amputation of a hopelessly damaged digit. Diabetes is responsible for as many as 60,000 amputations of toes each year.

Another region where capillary failure can have serious results is in the retina of the eye. Diabetes causes a seepage of blood from capillaries in the retina that can lead to blindness. Also, high levels of glucose in the fluids bathing the lens of the eye can cause the lens to become cloudier (a condition called cataracts).[5] All of these disturbing complications of diabetes make it clear that the most pressing problem for current-day endocrinology is finding out what is causing resistance to insulin in type II diabetes. If only this puzzle could be solved, it might be possible to find a cure for type II diabetes.

One helpful clue is that the development of obesity is often accompanied by an increasing resistance of the body to insulin. But how would fat cells cause insulin resistance? Most (85%) of the insulin-stimulated glucose uptake in the body occurs in muscle cells and *not* in fat cells.[30] How could fat cells cause muscle cells to change their responses to insulin?

Researchers have focused on a number of molecules released from fat cells that may influence the responses of muscles to insulin. Some of these, like free fatty acids and newly discovered protein molecules, do seem to have some effects on the responses of muscles to insulin. But one fat cell protein in particular that may be of critical importance is our old friend, the fat cell hormone leptin.

In 2004, Timo Lakka and coworkers at the Pennington Biomedical Research Center in Baton Rouge, Louisiana, performed a study to see if small differences in the structures of leptin or of leptin receptors could have any effect upon overall glucose metabolism. It has been known for a few years that there are three slightly different types of leptin receptors in the general population. In one of these subtypes, it is common to inherit a tiny change in the gene that can be detected with specialized DNA analysis techniques. Lakka gathered information on these subtle differences in the genes for leptin receptors from 397 adults and then subjected them to an aerobic exercise program in the gym for 20 weeks. After the training period, many of the people showed great improvements in overall insulin sensitivity and had lowered levels of blood glucose. This was expected, since it has long been known that exercise and dieting can greatly improve the condition of diabetics by increasing the responsiveness of muscles to insulin. However, to Lakka's surprise, a subset of people with a particular subtype of leptin receptor failed to show any improvement at all! Since almost 50% of the general population inherits this "bad" form of leptin receptor, it could be that a failure of leptin's action is a major player in the development of diabetes in the United States.[23]

How would leptin affect the insulin sensitivity of muscles? Leptin receptors are present on muscle cells, but when leptin is applied to muscle cells that have been isolated and grown in a dish, no direct effects of leptin on glucose uptake by muscle cells can be seen![50] So, where does leptin act to modulate the insulin sensitivity of the body, and how does this action relate to diabetes mellitus?

Joel Elmquist and coworkers at Harvard University recently completed a series of ingenious experiments to test the idea that leptin-sensitive neurons of the arcuate nucleus could be key regulators of overall glucose metabolism. They started out by creating a genetically engineered mouse that synthesized an abnormal form of the leptin receptor in all the cells of the body; these mice, consequently, were fat, resistant to leptin, and resistant to insulin. These findings in themselves were not so new—after all, it has long been known that genetically obese mice have an abnormal form of the leptin receptor and have an abnormal glucose metabolism. What was novel in Elmquist's experiments was that the abnormality in the mouse leptin receptor could easily be *reversed* and *completely corrected* by simply exposing cells to an enzyme! How was this done?

The creation of reversible resistance to leptin in a mouse was accomplished by manipulating the DNA inside the cells of an early mouse embryo. The cells of these mice were genetically engineered to contain extra DNA within the gene for the leptin receptor. This extra DNA contained sequences that, when translated into a protein, produced a nonworking form of the leptin receptor. These extra DNA sequences also contained sites that are recognized by a special enzyme produced by yeast cells. This enzyme, called a "flippase recombination protein," settles down onto DNA, binds to specific sites (flippase recombination targets, or FTPs), and then cuts a large loop out of the DNA and excises the offending sequences from the gene. So, if a cell could simply be exposed to this enzyme, its gene for the leptin receptor and its ability to respond to leptin could be easily and completely corrected.

Elmquist and his colleagues were not simply interested in restoring responsiveness to leptin in all the cells of the body. They wanted to see if restoring responsiveness to leptin in the arcuate nucleus of the hypothalamus alone would be sufficient to restore a mouse to normal health. This posed an additional challenge.

As we have seen in previous pages, it is relatively simple to introduce chemicals into the brains of mice. Normally, things like leptin, glucose, or neurotransmitters can be infused into the hypothalamus via small metal tubes called cannulae. These were really not good enough for the aims of Elmquist and coworkers, however. Metal cannulae typically have a diameter of about 1 millimeter. This sounds pretty small, but the arcuate nucleus of a mouse is only about one-half of a millimeter wide! Attempting to introduce a chemical into such a small arcuate nucleus with a metal cannula is rather like trying to water your garden with a sewer pipe instead of a hose. A metal cannula is just too large and would flood most of the bottom of the brain with the flippase enzyme. To avoid this problem, a much smaller tube would have to be used to specifically target the arcuate nucleus and only the arcuate nucleus. Where can one find such a tiny tube?

Fortunately, really tiny tubes are readily available to neuroscientists. These tubes are called glass micropipettes. These have diameters of less than 1/100 of a millimeter and can deliver very tiny amounts of fluid into very small spaces. These delicate properties make them sound like very sophisticated devices, but in actual fact I was able to make a couple of them while working with collaborators in our Department of Physiology, and if I can make them, I can assure you that you could make them, too. All you need are two things: (1) small, hollow glass tubes called capillary tubes (these are commonly used to collect blood from fingertip punctures in the doctor's office and have diameters of a couple of millimeters) and (2) a micropipette puller.

A micropipette puller is a simple machine. To use it, you place one end of a vertically positioned capillary tube into a metal clamp and then attach a small weight to the other end of the tube. To convert the capillary tube into a glass micropipette, a thin coil of wire is wrapped around the middle of the capillary tube and is then electrically heated. As the glass in the capillary tube heats up, it starts to soften and stretch just like a piece of saltwater taffy. Finally, in response to the heat and the weight at the end of the capillary tube, the glass is pulled into a long and very fine process that still contains a hollow cavity. Such a delicate tube is perfect for infusing the flippase enzyme into small clusters of cells in the hypothalamus.

When the flippase enzyme was infused into the arcuate nucleus of genetically engineered mice, the neurons there converted their abnormal genes into a normal gene for the leptin receptor and began making new, functional leptin receptors for themselves. Within four weeks of this procedure, the mice had greatly improved sensitivity to insulin and started to recover normal levels of blood insulin and glucose, even before they lost significant amounts of body weight.[8] Judging from these experiments, it seems likely that leptin-sensitive neurons in the arcuate nucleus of the hypothalamus have potent effects on overall glucose metabolism, probably by altering the function of sympathetic nerve endings that synapse onto muscle cells.

A more recent study by the same group showed that intrahypothalamic infusions of leptin are also very beneficial to mice with a form of type I diabetes (these mice were made diabetic by damaging the pancreatic beta cells with streptozotocin, a chemical that causes beta cell death in a manner very similar to that caused by alloxan).[14] Both of these studies suggest that diabetes may not only be a disease of the pancreas but might also be a disease provoked by abnormal hypothalamic function. The quest for a treatment that would restore normal function of leptin in the hypothalamus would thus not only be beneficial in treating obesity but would probably also be very helpful in treating diabetes.

AGING, HORMONES, CANCER, AND THE HYPOTHALAMUS

Aging causes most of the organ systems of the body to deteriorate in some ways, and the endocrine system is no exception. Changes in the control of blood glucose by insulin and other hormones are probably the most notable and potentially the most damaging of these age-related changes. After the age of 65, about 40% of us are destined to show insulin resistance, impairments in glucose metabolism, and elevated blood levels of insulin that may progress to diabetes mellitus.[24] How much the hypothalamus contributes to this disagreeable change is not known. Aging of the pancreatic islets themselves could also contribute to this problem.

This age-related deterioration in sensitivity to insulin may not only provoke diabetes but now seems related to other serious problems that develop with age. Cancer, surprisingly, is one of these problems. How could the aging of the hypothalamus possibly contribute to the risk for cancer?

Cancer has been an aging-related disease that humans have struggled with for some time. The vast majority of cancers occur in people who are older than 50 years or so. For example, the overall incidence of colorectal cancer is less than 10 per 100,000 people aged 40 or less but zooms to 300 per 100,000 people by age 70. So, in many ways, cancer is a disease of aging. Many theories have tried to explain this link between aging and cancer, and there are plenty of potential explanations. One probable one is that damage to chromosomes or mutations in DNA that provoke cancer take a long time to occur and accumulate in numbers sufficient to disable a cell. Another possible explanation is that the immune system deteriorates with aging and is no longer so able to detect and destroy cancer cells.

In 1971, President Nixon prompted the creation of new research funding for a so-called war on cancer that was aimed toward curing cancer within a decade or so. This war on cancer provided for thousands of studies of mechanisms of cell division that produced great advances in the basic understanding of cells. However, until relatively recently, these advances did not provide the benefit of reduced deaths from cancer. In fact, death rates from cancer in 1994 were actually 6% *higher* than death rates in 1970, in spite of all the advances in treatment and cancer biology![4] This might provoke the average taxpayer to complain, reasonably enough, that all that money spent on research had been wasted without any tangible progress in fighting cancer.

Fortunately, in the last five years or so, cell biologists *have* been able to identify molecules causing cancer and *have* been able to produce drugs that finally improve the cure rates for cancer. Most of these drugs target receptors on cell membranes that respond to growth-promoting molecules in the blood. Without the stimulation of these growth-promoting molecules, many cancer cells fail to grow well and thus finally can be successfully treated.[18] It turns out that one of the growth-promoting molecules in the blood is insulin.

One of the earliest researchers to point out an apparent link between diabetes, insulin resistance, and cancer susceptibility was a clinical endocrinologist from Leningrad named Vladimir Dilman, who observed increasing incidences of cancers in his diabetic patients.[11] More recently, this train of thought has been followed by other researchers. For example, a study of 1 million Americans, published in 2004, found that rates of different types of cancers were significantly higher in people with diabetes, insulin resistance, and higher-than-normal levels of insulin in the blood. Liver cancer, for example, is almost twice as frequent in diabetics as in normal patients; other cancers such as bladder cancer, colon cancer, pancreatic cancer, and breast cancer are also more common in diabetics.[9]

The probable explanation for a relationship between blood levels of insulin and cancer is that insulin binds not only to insulin receptors but also to receptors for insulin-like growth factor 1 (IGF-1) that stimulate cells to grow and multiply.[15] So, the aging-related increases in insulin resistance and blood insulin levels may not only signal disturbances in glucose metabolism but may also underlie the increased risk for cancer seen in aging. Once again, the contribution of hypothalamic dysfunction to all of this is still not known with certainty, even though a prominent role for the hypothalamus in the control of insulin sensitivity has been firmly established.

Hypothalamic dysfunction may also be related to another type of cancer, breast cancer. Breast cancer is the most frequently diagnosed type of cancer in the United States; about 200,000 new cases are reported every year and cause the death of about 40,000 people each year, a lethality second only to lung cancer. I have had some personal experience with this type of cancer. Fortunately, my experience involved observing tumors in rats rather than in people.

My unexpected introduction into cancer biology basically involved negligence on my part. In most of the experiments I have performed, I usually looked at how rats behaved in various ways and then gave them an overdose of anesthesia so that I could remove the brain and examine brain cells of various types in the hypothalamus. On a couple of occasions, I have completed the experiment and moved on to another one but neglected to count all my rats in the animal facility. After six months or more passed, I would look at the invoices from the animal facility and find that I was being charged extra for two or more rats. Then I would go up to our animal rooms, look through the cages, and find that the extra rats had been waiting for me all this time and had grown much larger and older. Old male rats can grow to a very large size, almost as big as squirrels (my graduate students would suggest that I could put a saddle on them and ride them around the room!). Female rats don't get this big, but they do develop yellowish patches in their previously white fur, and they also acquire mammary tumors that can grow to the size of a nickel. By the advanced age of two years, almost one-quarter of all female rats develop these tumors. Most of the tumors are benign and will not hurt the animal, but they are still interesting and have been used as a model system to study the properties of breast tumor cells. Why do old rats get these tumors so often?

The simple answer to this question is that dopamine-containing nerve cells in the hypothalamus of old rats seem to get tired and don't make as much dopamine as they should. Since dopamine normally inhibits the secretion of prolactin from the pituitary, blood levels of prolactin increase and promote the growth of tumors in mammary glands.

For years, this aging process and its associated development of mammary gland tumors was considered a peculiarity of the rat and wasn't thought to have any applicability to breast cancer in humans. Lately, however, this viewpoint has been discredited. It is now well established that higher blood levels of prolactin, due to hypothalamic aging, are associated with a twofold increase in the risk for developing breast cancer.[19] Drugs that block the effects of prolactin on breast tissue are now being studied to see if they could protect against this dangerous disease.

Cancer actually didn't used to be much of a health problem for us, simply because most of us never lived long enough to develop it! In 1900, the average life span in the United States was only 47 years, whereas by 1996 it had increased to 76 years.[13] Prior to industrialization, the development of procedures for maintaining clean water and noncontaminated food, and the discovery of antibiotics and vaccines, most of us succumbed to infectious diseases well before any tumors could develop.[11] This is still true in nonindustrialized, low-income countries: in 2008, 34% of the deaths in poor countries were from infections, but in rich countries, only 3.8% of deaths involved infections. Instead, cancer was a cause of 13% of all deaths in rich countries.[43] Thus, the conquest of the previous infectious causes of death has prolonged life, but it has also exposed us to new threats of abnormalities in physiology that seem to be partly connected to age-related changes in the function of the hypothalamus.

I actually had an opportunity to collaborate with Dr. Dilman in his studies showing links between insulin resistance and diseases of aging like cancer, all because I happen to work and live in the Washington, D.C., area. One of the benefits of my working here is that I am close to the huge research campus of the National Institutes of Health in nearby Bethesda, Maryland. For some time, I was on a mailing list from NIH that informed local scientists of visiting speakers who might be of interest to us. Dr. Dilman was invited to speak in 1991, and I took the subway from Howard University to the NIH to hear him. I got to the seminar a few minutes late, just in time to see him introduced by the organizer of the seminar, whose name I didn't hear. I found Dilman's seminar, delivered in highly accented English, to be very interesting, so I decided to approach him afterward and talk with him.

Now, I learned a little Russian in college and can read it pretty well, although speaking it is another matter altogether. However, I was able to communicate with Dr. Dilman well enough in my broken Russian, and he suggested that we go upstairs to the lab of the organizer of the seminar, who had invited him in the first place. I turned to this other scientist; introduced myself and absently heard his name, Dr. Gajdusek, in response; and shook his hand.

When I rode the elevator up several floors with Dr. Dilman and entered the lab of his host, I was puzzled to see large posters of New Guinea and other tropical locations on the walls. Then it struck me: didn't a man named Gajdusek publish important work on a neurological disease called kuru? As I thought about it some more, I realized I was in the lab of Dr. Carleton Gajdusek, who had won the Nobel Prize in 1976 for his work on viral infections of the nervous system in people of New Guinea. I was more than a little embarrassed that I hadn't recognized his name earlier, but he didn't seem to mind. Dr. Gajdusek was doing important work on mad cow disease, another infectious disease of the nervous system, at the time but was also interested in current Russian neuroscience because he also spoke the language fluently. Subsequently, his career faltered because of his illicit sexual activity with boys he had brought back to the United States from New Guinea. He served 12 months in jail for this offence and left the United States for Europe, never to return until his death in 2008. He was one of the few Nobel Prize winners whom I have had a chance to meet.

One good outcome of my meeting with Dr. Dilman was that I agreed to translate into English a book of his on his research. As I got to know him, I found him to be kind, good natured, and intellectually very ambitious, with wide-ranging ideas about aging and diseases. His book was one of the final things he and I accomplished before he died, ironically of cancer, in 1994.[11]

Cancer is not the only malady that can be worsened by high levels of insulin and by insulin resistance. Diseases of the coronary arteries that supply blood to the heart are two to four times more common in people with type

II diabetes mellitus.[42] Abnormal cardiac vessel function in diabetics is most likely due to general abnormalities in the metabolism of fats (lipids) that are seen in diabetes. Higher blood levels of lipids promote the deposition of plaques in coronary arteries. Also, insulin itself can cause abnormal activities of smooth muscle cells and the lining epithelial cells in blood vessels.

Heart disease is a leading cause of death in the United States (16.3% of all deaths). Other vascular disorders like stroke account for another 9.3% of deaths, and cancer (14% of deaths) and diabetes (2.8% of deaths) are also major contributors to mortality.[43] Since all of these disorders are aggravated by higher insulin levels and by insulin resistance, it's not unreasonable to claim that 42% of aging-related deaths are in some ways related to the development of insulin resistance. Thus, a better understanding of how the hypothalamus contributes to aging-related insulin resistance would seem to be potentially of great benefit in protecting people from diseases of aging.

THE HYPOTHALAMUS AND MENOPAUSE

Another drastic age-related change that is most dramatic in women is a loss of ovarian reproductive hormones that occurs in a gradual process called menopause. Menopause typically begins at about age 46 and lasts for about four years. During this time, monthly menstruation gradually ceases and women experience a number of peculiar symptoms. Any of us who have been close to our mothers, wives, or sisters at this time know very well that one such symptom is the appearance of sudden, wild variations in body temperature that are called "hot flashes" or "hot flushes." Suddenly, a woman undergoing menopause can feel unusually hot and will run over to an air conditioner or to a refrigerator for a cold drink. What is going on here?

Temperature changes during menopause are not at all subjective and can be studied in experimental animals. One ingenious approach to studying hot flashes was used by scientists at Fukuoka University in Japan. This team surgically removed the ovaries from mice and then, following their recovery, put them on treadmills for 10-minute bouts of exercise. The mice without ovaries developed hot, pink tails that had a higher-than-normal skin temperature. However, if the mice were treated with estrogen, they never developed these hot flashes. These scientists concluded that the removal of estrogen from the mice caused an alteration in the control of body temperature by the hypothalamus. Something similar must be going on in women undergoing menopause.[38] These ideas fit in with the ability of ovarian hormones to change body temperature earlier in life, as noted in chapter 2.

A disturbed control of body temperature is one annoying aspect of menopause, but it is relatively minor and transient compared with some other effects. A loss of ovarian estrogen during menopause can also decrease bone density and increase the risk for breaking bones. Also, women are about twice as likely as men to develop Alzheimer's disease when they get older, probably because of a loss of the neuron-protecting effects of estrogen.[31] So, menopause is not a trivial event in human life. What causes it?

If I were writing this book 10 years ago, it would have been easy for me to confidently answer this question. Ten years ago, everyone knew that the deterioration and final disappearance of egg cells (oocytes) from the ovaries was the obvious cause of menopause. Now, however, this simple explanation has suddenly become very controversial, and we no longer know for certain what is going on in the aging female reproductive system.

Why do ovarian egg cells disappear with age, and how would this affect the production of estrogens by the ovaries? Egg cells themselves do not make estrogen, after all. Anyone who wants to really understand menopause must first get to know a little bit about the ovaries.

Ovaries are accumulations of connective tissue that contain thousands of egg cells. These egg cells actually don't originate in the ovary at all. During embryogenesis, egg cells are found in the yolk sac of the embryo and migrate toward the developing ovaries in response to a chemical signal (stromal cell-derived factor) produced by the ovaries. Once each egg cell has found an ovary, it burrows into the surface of the ovary and makes itself at home.[2]

At birth, every baby girl will possess about 300,000 egg cells within each ovary. However, most of these egg cells are destined to either remain very small or disappear altogether. Only a few of them are mysteriously singled out for further development (and nobody has any clear idea how this works!). As these favored few egg cells enlarge, they stimulate the development of other ovarian cells nearby them that form enveloping structures called ovarian follicles.[33]

Over the course of 35 years, from puberty to menopause, most of the egg cells in the ovary degenerate, leaving behind a smaller and smaller pool of healthy egg cells that may ovulate and which maintain the structures of the surrounding hormone-secreting follicle cells. It has long been assumed that when the ovaries age and all the egg cells are gone, the follicles disappear and their hormones disappear with them. This assumption now appears to be wrong.

For the last five years or so, Jonathan Tilly and coworkers at Harvard Medical School have come up with evidence that the prevailing view of ovarian aging is incorrect. They have found stem cells in ovaries from very old (24-month-old) mice that have many of the proteins normally found only in egg cells. In old mice, these stem cells fail to fully mature into egg cells, but if an ovary from an old mouse is transplanted into the abdominal cavity of a young (2-month-old) mouse, these stem cells "wake up," enlarge, and form new egg cells surrounded by follicles. In contrast, if ovaries from a young mouse are transplanted into an old mouse, the stem cells "go to sleep" again and fail to form new egg cells.[29] These experiments seem to show that the ovaries themselves are not to blame for reproductive aging in mice; instead, some factors in the body of an old mouse that suppress the ability of an ovary to stay young must be responsible for reproductive aging.

What might these factors be? A leading researcher of reproductive aging, Phyllis Wise at the University of Washington in Seattle, has identified a number of changes in the hypothalami of aging rodents that could explain menopause in rats. She has found that the LH-RH containing neurons of the hypothalamus show diminished reactions to estrogen in old rats. This causes a diminished ability to produce a surge in the secretion of LH from the pituitary that causes ovulation.[12] It seems that abnormalities in the control of reproductive hormones by the aging hypothalamus may be an underlying cause of reproductive aging in rats.

Do any of these data apply to humans? If the aging hypothalamus could be "kick-started" somehow, would this normalize blood levels of reproductive hormones and reactivate ovarian stem cells in humans? A reversal of the aging of the female reproductive system would be a highly desirable goal. However, while ovarian stem cells do appear to also be present in human ovaries, it is not clear if they could be used to combat aging. Blood levels of LH in humans do not show the same types of changes that are seen in aging rats. Moreover, estrogen-sensitive neurons in the human hypothalamus seem to become *more* active with aging, rather than less active as in rats.[34] So, data from human subjects may put the spotlight on ovarian problems rather than hypothalamic problems during aging. More research on this topic clearly needs to be done.

AGING OF HYPOTHALAMIC CELLS

One peculiar sign of aging in the hypothalamus of both rats and humans is the appearance of a strange type of glial cell (a subtype called a Gomori-positive astrocyte). Beginning at about two months of age in rats and mice, these astrocytes start to develop dark-staining granules in their cytoplasms that can be stained using a special stain called the Gomori stain. The presence of these aging-associated cells has been known for at least 50 years, but no one knew where the astrocyte granules came from or why they formed.

Dr. Hyman Schipper in Montreal has been one of the most active investigators of these peculiar cells. In 1994, Dr. Schipper and coworkers closely examined the development of these Gomori-positive astrocytes using an electron microscope and found that their mysterious granules were formed when mitochondria in these cells degenerated; the cytoplasmic granules are remnants of mitochondrial debris and metal-containing proteins. The aging process somehow seems to damage the mitochondria of these cells. They are by far most abundant in the arcuate nucleus of the hypothalamus, but they can also be found in smaller numbers in other brain regions of rats, and also of humans (Fig. 5-1)![36]

What could be damaging the mitochondria of glial cells in the hypothalamus? And why does the same mysterious aging-linked process not also damage the mitochondria of neurons right next to them? These kinds of questions have puzzled me and recently have been the focus of my own research. Dr. Schipper and I have met at scientific meetings over the years and write each other occasionally about these mysteries, but neither of us feels that we have found all the answers yet.

One possible explanation for this peculiar cellular pathology is that the arcuate nucleus is continually exposed to harmful substances that percolate out of the leaky capillaries in this area. Perhaps some of these substances may be metabolized by mitochondria and provoke oxidative stress that damages these organelles. I have found that Gomori-positive astrocytes are unusually enriched in proteins that bind lipid molecules like fatty acids.[47] It turns out that the mitochondria of neurons cannot metabolize fatty acids, but for unknown reasons, the mitochondria of astrocytes can burn these molecules.[37] So, perhaps the continual burning of fatty acids by glial mitochondria could be one explanation for why these cells develop mitochondrial damage.

It is not known for sure if damaged mitochondria in astrocytes could contribute to aging-related disturbances in neuronal function. However, in mice genetically engineered to produce an abnormal mitochondrial enzyme only in astrocytes, the overall function of the entire nervous system is damaged.[44] So it is possible that the gradual development of mitochondrial abnormalities in Gomori-positive astrocytes in the hypothalamus may be a warning sign that aging is damaging the function of this important brain region.

Some years ago, I showed that Gomori-positive astrocytes are often in close contact with dopaminergic neurons of the arcuate nucleus.[49] These neurons have a number of functions. For example, they inhibit the release of prolactin from the pituitary gland. If aging damages the function of adjacent Gomori-positive astrocytes, this may be one reason why dopaminergic neurons show abnormalities with aging and fail to restrain prolactin secretion, thereby contributing to the development of mammary gland cancer.

Dopaminergic neurons also are activated during hypoglycemia and seem to regulate blood glucose.[6] The Gomori-positive astrocytes next to these

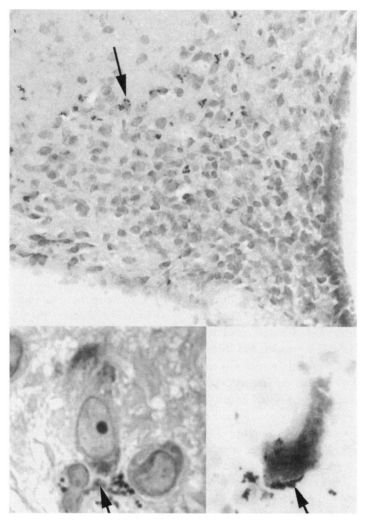

Figure 5.1. Top: Overview of the arcuate nucleus of an old mouse, showing the locations of astrocytes with degenerating mitochondria (arrows). Bottom right: Rat hypothalamus, showing a Gomori-positive astrocyte with numerous cytoplasmic granules derived from aging mitochondria (arrow). Left: Human hypothalamus, showing a Gomori-positive astrocyte in close contact with a dopaminergic neuron that contains accumulations of neuromelanin in the cytoplasm.

neurons possess the glucose-sensing GLUT2 type of glucose transporters and thus may help the dopaminergic neurons detect and respond to changes in blood glucose.[48] Any aging-related changes in the functions of these neurons and their associated astrocytes could thus lead to aging-related abnormalities in the control of glucose metabolism by the hypothalamus.

Finally, Gomori-positive astrocytes also frequently contact leptin-sensitive neurons and potentially could also influence their function of restraining appetite.[45] Thus, these poorly understood astrocytes seem worthy of more study in the context of aging-related dysfunction of the hypothalamus.

· 6 ·

Some Final Thoughts on Mice and Men

$\mathcal{A}$fter reading through the first five chapters of this book, you must have concluded that much of what we know about the hypothalamus comes from experiments on animals. To be sure, many of the disorders of the hypothalamus that attracted interest to it in the first place, such as acromegaly or Frölich's syndrome, were first identified in humans, but a final understanding of how the hypothalamus could cause these human syndromes ultimately came from experiments on animals. The use of animals in research can be criticized as just another example of how humans exploit animals for their own selfish purposes. Researchers, moreover, can be accused of confining animals in inhumane research cages and exposing them to discomfort and death to conduct their research.

There is no denying that these accusations are in some ways true. We *do* utilize animals for our own selfish purposes, and these purposes can involve discomfort for animals that they are helpless to avoid.

It doesn't make me happy to feel that I might be treating animals in a cruel or selfish way. I enjoy the company of animals as much as anyone else. My wife and I maintain a large back yard, with a bird feeder that attracts cardinals, goldfinches, and wrens; occasionally we get to know some of these birds as individuals because of missing feathers or an injury that marks them out as specific, identifiable neighbors. We also have other visitors to our yard like chipmunks, squirrels, turtles, foxes (rarely), and even an occasional deer. I am always delighted to see these beautiful creatures.

When these wild animals come into our house, I have to admit that my attitude changes somewhat. Every once in a while, field mice squeeze through some tiny cracks in our house and run around the basement, chewing on everything and leaving their droppings. Occasionally, they even make their way, unseen at night, up into our kitchen. At this point I never hesitate to get out mousetraps that eliminate this challenge to our domestic sovereignty. Even so, it is not a happy moment for me to see the life of even these tiny mice come to an end. The same thing applies to my work in the lab: in order to examine brain cells of the hypothalamus, I have to inject mice and rats with overdoses of anesthetic. It is never a pleasant experience to see the light of life fade from their eyes before I open up the skull to remove the brain.

Should I and other researchers feel guilty and immoral because I use rats and mice in my research? There may not be a straightforward answer to this question. Perhaps putting the use of animals in research in a broader context may be helpful.

In 1998, about 23 million mammals were used in research, along with smaller numbers of amphibians, fish, and insects like fruit flies. About 95% of the mammals in research are mice and rats, while 76,000 dogs, 25,000 cats, and 57,000 primates make up the remaining 5% of mammals.[2] Most people focus on the controversy of using cats and dogs in research, probably because their experiences with these animals as pets makes them aware of the emotional and intellectual qualities of these animals. I myself tend not to make so much of a distinction between dogs and rats and mice. It's true that smaller-brained animals probably have a less sophisticated life experience than larger ones, but it seems to me that emotions and intelligence are not confined to dogs and cats.

An outstanding example of the surprising abilities of animals can be found in the story of an African grey parrot named Alex who was studied by a researcher named Irene Pepperberg for 30 years. Dr. Pepperberg found that Alex could understand concepts like color and shape, allowing him to distinguish the differences between gold keys and blue keys, and could accumulate a vocabulary of over 100 words. Alex displayed an amazing range of intellectual abilities. When Alex unexpectedly died, Dr. Pepperberg felt a genuine loss at losing a loved one with whom she had formed a close relationship.[4] If a bird can demonstrate a surprising intellectual and emotional degree of sophistication, it seems likely to me that many other animals have these traits as well. The brain of a rat is not so different in size from the brain of a parrot.

Even if a rat had a blunted ability to reason or to feel emotion, would that be a sufficient rationale for using the rat for selfish human purposes? Are only intelligent animals valuable? One reason why many of us feel upset about the practice of harvesting whales from the sea is that these animals are demonstrably intelligent and have complex songs that allow them to communicate with each other. However, cows also have large brains and are undoubtedly intelligent in their own right, but that doesn't stop us from killing and eating them.

Should only relatively less intelligent animals be used in research, so that more intelligent animals like monkeys should be spared? Where do we draw the line? What would be a critical level of intelligence that would permit research on an animal? How would this level of intelligence be measured?

What about humans? Should more intelligent humans have more rights than less intelligent humans? I would guess that most people would say no to this proposal. If so, then the same principle would suggest that intelligence might not be a decisive factor in deciding whether or not a particular species of animal should be employed in research. In an absolute sense, all animals have some intelligence, so perhaps no animals, including fruit flies or worms, could ethically be used for research.

These thorny questions get even more complicated when we consider other ways that humans use animals. By far the most extensive use of animals by humans occurs in the food industry. Each year, about 9 billion chickens, 100 million pigs, 250 million turkeys, and 36 million cows are slaughtered for food for humans.[1] I'm sure that if all these animals had a choice, they would rather not be raised for eating by humans.

The keeping of animals as pets also intrudes upon the free life of an animal. Even if pets have relatively happy lives, in order to feed them dog or cat food, other animals have to be slaughtered to provide the meat. Moreover, since many people fail to spay or neuter their pets, dogs and cats reproduce at much higher rates than are desirable, so that humans have to deal with many unwanted pets. Every year, 3 to 7 million unwanted pets are euthanized (mainly by inhalation of carbon dioxide) in animal pounds. This number is far higher than the number of dogs and cats used in research. People concerned with the rights of animals could also justifiably criticize the keeping of animals as pets as well as the use of animals in research.

Now, it could be argued that, since every organism dies, there is no real harm in using the body of a dead animal for food, particularly if the animal experiences a relatively pleasant life before dying. Studies of human anatomy and physiology show that we have characteristics that we share with other meat-eating animals like tigers (e.g., we have canine teeth rather than the grinding teeth of herbivores, and our gastrointestinal system is adapted for the digestion of meat). I'm sure that a tiger feels no guilt when he attacks and devours a prey animal. Should we, then, feel guilty when we eat meat, particularly when the life of a food animal prior to death was relatively good?

Lives of animals under human supervision can vary considerably. I would have to say that the life of a dairy cow, for example, seems to me to be pretty good. Each cow is given free reign of a pasture, has all the grass to eat that it wants, and has good shelter from bad weather and from predators, all in exchange for a daily delivery of milk. Sounds like a pretty good deal to me in comparison with the hazards of a life in the wild. The lives of other food animals, confined within feed lots (pigs and cattle) or in small cages (chickens and turkeys), may not be so desirable and perhaps could be improved if humans did not mind paying more for their upkeep.

It seems to me that anyone wanting to live a life that does not hurt animals must first focus on the eating of meat, which affects far greater numbers of animals than does research. I have considered taking up a vegetarian diet for just such a reason but have always personally found it to be very hard. It's not that I don't like vegetables—my wife and I eat cauliflower, broccoli, carrots, and other vegetables all the time and enjoy them. It's also not that I'm a lousy cook (though this is one argument that could be made!). I have consulted vegan cookbooks in search of vegetarian meals that don't utilize animal flesh. The problem is that it is not so easy to come up with all-vegetable meals that contain sufficient protein to maintain healthiness. I remember one time that I cooked up one concoction of squash and tomatoes that was supposed to substitute for a mainstream meal. We both took a couple of spoonfuls of it, found it extremely distasteful (even the texture of it was bad), threw it out, and wound up having fried eggs and toast for dinner! I have just not been able to give up meat or fish in my diet, to say nothing of cheese, milk, or eggs (or even noodles, which contain eggs, or soups, which contain chicken stock). Personally, I find it very difficult to live a life that does not intrude upon the lives of other animals. Who knows, though; perhaps I will find a vegan menu that I can get used to and someday will be able to eliminate or reduce my consumption of meat.

Perhaps the greatest harm that humans do to animals worldwide comes from the overharvesting of wild creatures, the replacement of their habitats with farms and communities, and the burning of fossil fuels that foster global warming. It is estimated that overfishing of the seas caused the total biomass of predatory fishes to decline by almost two-thirds during the twentieth century, leading to constantly declining fish catches over the last few decades.[3] Also, by leaving the lights on in office buildings overnight, we cause the deaths of millions of birds annually when they crash into glass windows. All of us could probably help wildlife in some ways. One solution to these man-made injuries is simply to lower the rate of human population growth so that we can share the planet with other animals more equitably. Research on the hypothalamus and endocrine system has provided one way to reduce the population explosion in the form of birth control methods, so animal research may be part of the solution rather than part of the problem.

Even if research impacts the lives of animals much less than the food industry, couldn't it still be argued that surgical "vivisection" on animals is a cruel and unnecessary practice? It is true that some of the surgical procedures used on animals during the early days of neuroscience would probably not be allowed now because they inflicted too much pain and suffering. Nowadays, however, the rules for researchers are much stricter.

In order to receive funding from public or private sources of research money, every researcher now must submit a research proposal to a committee at his university called an institutional animal care and use committee (IA-CUC). Each committee is composed of both scientists and nonscientists, and all of the experiments in a given proposal are closely scrutinized to ensure that the proposed work (1) does not duplicate work that has been carried out by others, (2) uses the minimum number of animals needed, and (3) minimizes the imposition of pain upon an animal. For example, I'm allowed under these rules to give an injection to an animal without any restrictions, but if I need to make an incision into the skin of an animal for any reason, I not only have to utilize a general anesthetic and sterile techniques but also have to provide painkillers in the drinking water of the animal for the five to six days it takes for the incision to heal. This is the same type of treatment that doctors are required to apply to humans.

I also have to demonstrate that I have been trained to pick up and handle an animal in ways that do not stress or frighten it. Rats and I have gotten to be good friends over the years, and we have become so comfortable with each other that a rat can roam freely around my lab coat without either of us harming the other. Mice, on the other hand, are not as nice to me as rats and will aggressively bite each other, me, and anyone else in the lab if given the opportunity, so I'm a little more wary of mice.

Other regulations ensure that the lives of animals at research institutions are comfortable. Animal research facilities have to pass rigorous inspections every year. Each facility must maintain comfortable temperatures for the animals, all cages must be regularly steam cleaned to reduce possibilities for infections, and fresh food and water must be continuously available. It is true that rats and mice must live in cages, but it is useful to remember that these are burrowing animals that prefer to live close to each other in small spaces. If rats are introduced into a large space, they will huddle together in a corner to avoid crossing the floor and entering an open space that they instinctively associate with dangers from predators in the wild. So, life in a cage may not be as much of a hardship for them as one might at first think.

The use of dogs and cats in research makes me a little bit more uncomfortable, even though there is no doubt that some scientific and surgical problems could never have been overcome without them. Confining dogs who were born to roam across large areas in a small research space appears to place restrictions on them that seem inhumane to me. Similarly, it may not be such a great idea to keep pet dogs confined to small apartments or homes for most of the day, or to keep large animals confined in zoos. It is hard to know where to draw the line in such cases. Is the keeping of a pet bird in a cage just as unethical as keeping a mouse in a lab? As long as humans have domain over other animals, due to our ability to manipulate them, we have to be somewhat concerned about using our powers ethically.

In the final analysis, if we are being honest with ourselves, the maintenance of the life of any human or animal inevitably involves some harm to the life of other organisms (even cows must destroy the lives of the plants that they eat). I'm not sure if any of us could reasonably hope to live out our lives without in some way contributing to the harm of other animals. If avoiding harm to animals is to be a legitimate goal, then a reduction in the human population itself seems the most reasonable way to attain that goal. The use of animals in research seems to me to be just as ethical as other human uses for animals, as long as cruel procedures are not part of the research. Animal research, by leading to numerous treatments for endocrine disorders, brain disorders, cancer, and other diseases, has immeasurably benefited humans. It cannot be discarded without also discarding the hope of continuing to keep humans healthy.

Notes

INTRODUCTION

1. Kiani R, et al. 2007. Object category structure in response patterns of neuronal population in monkey inferior temporal cortex. *J. Neurophysiol.* 97:4296–4309.

2. Quian Quiroga R, et al. 2009. Explicit encoding of multimodal percepts by single neurons in the human brain. *Curr. Biol.* 19:1308–1313.

CHAPTER 1

1. Benoit SC, et al. 2009. Palmitic acid mediates hypothalamic insulin resistance by altering PKC-θ subcellular localization in rodents. *J. Clin. Investigation* 119:2577–2589.

2. Bugarith K, et al. 2005. Basomedial hypothalamic injections of neuropeptide Y conjugated to saporin selectively disrupt hypothalamic controls of food intake. *Endocrinology* 146:1179–1191.

3. Chicurel M. 2000. Whatever happened to leptin? *Nature* 404:538–540.

4. Ciofi P, et al. 2009. Brain-endocrine interactions: a microvascular route in the mediobasal hypothalamus. *Endocrinology* 150:5509–5519.

5. Ding F, et al. 2008. SnoRNA *Snord116 (Pwcr1/MBII-85)* deletion causes growth deficiency and hyperphagia in mice. *PLoS ONE* 3(3):e1709. doi:10.1371/journal.pone.0001709

6. Donoso MA, et al. 2010. Increased circulating adiponectin levels and decreased leptin/soluble leptin receptor ratio throughout puberty in female ballet dancers: association with body composition and the delay in puberty. *Eur. J. Endocrinol.* 162:905–911.

7. Gibson WT, et al. 2004. Congenital leptin deficiency due to homozygosity for the Delta133G mutation: report of another case and evaluation of response to four years of leptin therapy. *J. Clin. Endocrinol Metab.* 89:4821–4826.

8. Gregg EW, et al. 2010. Estimated county-level prevalence of diabetes and obesity—United States, 2007. *J. Am. Med. Assoc.* 303:933–935.

9. Grodstein F, et al. 1996. Three-year follow-up of participants in a commercial weight loss program. *Arch. Intern. Med.* 156:1302–1305.

10. Herculano-Housel S, Lent R. 2005. Isotropic fractionator: a simple, rapid method for the quantification of total cell and neuron numbers in the brain. *J. Neuroscience* 25:2518–2521.

11. Hill JO, et al. 2003. Obesity and the environment: where do we go from here? *Science* 299:853–855.

12. Homma A, et al. 2006. Differential response of arcuate proopiomelanocortin- and neuropeptide Y-containing neurons to the lesion produced by gold thioglucose administration. *J. Comp. Neurol.* 499:120–131.

13. Houseknecht KL, et al. 1998. The biology of leptin: a review. *J. Anim. Sci.* 76:1405–1420.

14. Karapanou O, Papadimitriou A. 2010. Determinants of menarche. *Reprod. Biol. Endocrinol.* 8:115–120.

15. Kublaoui BM, et al. 2008. Oxytocin deficiency mediates hyperphagic obesity of *Sim1* haploinsufficient mice. *Mol. Endocrinol.* 22:1723–1734.

16. Levine JA, et al. 2005. Interindividual variation in posture allocation: possible role in human obesity. *Science* 307:584–586.

17. Lu SC, et al. 2009. An acyl-ghrelin-specific neutralizing antibody inhibits the acute ghrelin-mediated orexigenic effects in mice. *Mol. Pharm.* 75:901–907.

18. Merchant AT, et al. 2009. Carbohydrate intake and overweight and obesity among healthy adults. *J. Am. Dietetic Assoc.* 109:1165–1172.

19. Morgane PJ. 1979. Historical and modern concepts of hypothalamic organization and function. In: Morgane PJ, Panksepp J, eds. *Handbook of the Hypothalamus.* Vol. 1, *Anatomy of the Hypothalamus.* New York: Marcel Dekker; 1–57.

20. Ness-Abramof R, Apovian CM. 2006. Diet modification for treatment and prevention of obesity. *Endocrine* 29:5–9.

21. Rodriguez EM, et al. 2010. The design of barriers in the hypothalamus allows the median eminence and the arcuate nucleus to enjoy private milieus: the former opens to the portal blood and the latter to the cerebrospinal fluid. *Peptides* 31:757–776.

22. Santollo J, Eckel LA. 2008. Estradiol decreases the orexigenic effect of neuropeptide Y, but not agouti-related protein, in ovariectomized rats. *Behav. Brain Res.* 191:173–177.

23. Schaller F, et al. 2010. A single postnatal injection of oxytocin rescues the lethal feeding behavior in mouse newborns deficient for the imprinted Magel2 gene. *Human Molec. Genetics* 19:4895–4905.

24. Scherag S, Hebebrand J, Hinney A. 2010. Eating disorders: the current status of molecular genetic research. *Eur. Child. Adolesc. Psychiatry* 19:211–226.

25. Schweiger BM, et al. 2011. Menarche delay and menstrual irregularities persist in adolescents with type 1 diabetes. *Reprod. Biol. Endocrinol.* 9:61.

26. Seo S, et al. 2000. Requirement of Bardet-Biedl syndrome proteins for leptin receptor signaling. *Human Molec. Genet.* 18:1323–1331.

27. Södersten P, et al. 2008. Behavioral neuroendocrinology and treatment of anorexia nervosa. *Frontiers Neuroendocrinol.* 29:445–462.

28. Soriano-Guillen L, Argente J. 2011. Central precocious puberty: epidemiological, etiological, and diagnostic-therapeutic aspects. *An. Pediatric.* 74:336–343.

29. Tennese AA, Wevrick R. 2011. Impaired hypothalamic regulation of endocrine function and delayed counterregulatory response to hypoglycemia in Magel2-null mice. *Neuroendocrinology* 152:967–978.

30. Versini A, et al. 2010. Estrogen receptor 1 gene (ESR1) is associated with restrictive anorexia nervosa. *Neuropsychopharmacology* 35:1818–1825.

31. Willer CJ, et al. 2009. Six new loci associated with body mass index highlight a neuronal influence on body weight regulation. *Nature Genetics* 41:25–34.

32. Yan J-W, et al. 2005. Proteome analysis of mouse primary astrocytes. *Neurochem. Int.* 47:159–172.

33. Young JK. 1986. Thyroxine treatment reduces the anorectic effect of estradiol in rats. *Behav. Neurosci.* 100:284–287.

34. Young JK. 2010. Anorexia nervosa and estrogen. Current status of the hypothesis. *Neurosci. Biobehav. Rev.* 34:1195–1200.

35. Young JK. 2010. *Introduction to Cell Biology.* London: World Scientific Press; 159.

36. Young JK, McKenzie JC. 2004. GLUT2 immunoreactivity in Gomori-positive astrocytes of the hypothalamus. *J. Histochem. Cytochem.* 52:1519–1524.

37. Young JK, Polston E. 2012. Specialized features of the arcuate nucleus of the hypothalamus. In: Dudas B, ed. *The Human Hypothalamus: Anatomy, Functions, and Disorders.* New York: Nova Science Publishers.

38. Young JK, Stanton GB. 1994. A three-dimensional reconstruction of the human hypothalamus. *Brain Res. Bullet.* 35:323–327.

CHAPTER 2

1. Bicego KC, et al. 2007. Physiology of temperature regulation: comparative aspects. *Comp. Biochem. Physiol. Part A* 147:616–639.

2. Conti B, et al. 2006. Transgenic mice with a reduced core body temperature have an increased lifespan. Science 314:825–828.

3. Drew KL, et al. 2007. Central nervous system regulation of mammalian hibernation: implications for metabolic suppression and ischemia tolerance. *J. Neurochem.* 102:1713–1726.

4. Duan P-G, Kawano H, Masuko S. 2008. Collateral projections from the subfornical organ to the median preoptic nucleus and paraventricular hypothalamic nucleus in the rat. *Brain Res.* 1198:68–72.

5. Fitzsimons JT. 1998. Angiotensin, thirst, and sodium appetite. *Physiol. Reviews* 78:585–667.

6. Gavrilova O, et al. 1999. Torpor in mice is induced by both leptin-dependent and -independent mechanisms. *Proc. Natl. Acad. Sci. USA* 96:14623–14628.

7. Grossberg AJ, et al. 2010. Arcuate nucleus proopiomelanocortin neurons mediate the acute anorectic actions of leukemia inhibitory factor via gp130. *Endocrinology* 151:606–616.

8. Grossberg AJ, et al. 2011. Inflammation-induced lethargy is mediated by suppression of orexin neuron activity. *J. Neuroscience* 31:11376–11386.

9. Hanada R, et al. 2009. Central control of fever and female body temperature by RANKL/RANK. *Nature* 462:505–509.

10. Harding AJ, et al. 1996. Loss of vasopressin-immunoreactive neurons in alcoholics is dose-related and time-dependent. *Neuroscience* 72:699–708.

11. Kondo N, et al. 2006. Circannual control of hibernation by HP complex in the brain. *Cell* 125:161–172.

12. Krause M, et al. 2010. A pause in nucleus accumbens neuron firing is required to initiate and maintain feeding. *J. Neuroscience* 30:4746–4756.

13. McKinley MJ, et al. 2006. Water intake and the neural correlates of the consciousness of thirst. *Semin. Nephrol.* 26:249–257.

14. Millan EZ, et al. 2010. Accumbens shell-hypothalamus interactions mediate extinction of alcohol seeking. *J. Neuroscience* 30:4626–4635.

15. Morrison SF, et al. 2008. Central control of thermogenesis in mammals. *Exp. Physiol.* 93:773–797.

16. Naeini RS, et al. 2006. An N-terminal variant of Trpv1 channel is required for osmosensory transduction. *Nature Neuroscience* 9:93–97.

17. Navarro M, et al. 2009. Deletion of agouti-related protein blunts ethanol self-administration and binge-like drinking in mice. *Genes Brain Behav.* 8:450–458.

18. Pelz KM, et al. 2008. Monosodium glutamate-induced arcuate nucleus damage affects both natural torpor and 2DG-induced torpor-like hypothermia in Siberian hamsters. *Am. J. Physiol. Regul. Integr. Comp. Physiol.* 294:R255–R265.

19. Phillips MI, Schmidt-Ott KM. 1999. The discovery of renin 100 years ago. *News Physiol. Sci.* 14:271–274.

20. Shimizu H, et al. 2007. Glial Na_x channels control lactate signaling to neurons for brain [Na+] sensing. *Neuron* 54:59–72.

21. Troebst CC. 1965. *The Art of Survival.* New York: Doubleday.

22. Walter I, Seebacher F. 2009. Endothermy in birds: underlying molecular mechanisms. *J. Exp. Biol.* 212:2328–2336.

23. Young JK. 2010. Glial cells, the unsung heroes of the brain. In: Young JK, *Introduction to Cell Biology.* Singapore: World Scientific Press.

24. Young JK, Grizard J. 1985. Sensitivity to satiating and taste qualities of glucose in obese Zucker rats. *Physiol. Behav.* 34:415–421.

CHAPTER 3

1. Allison T, et al. 1996. Localization of functional regions of human mesial cortex by somatosensory evoked potential recording and by cortical stimulation. *Electroenceph. Clin. Neurophysiol.* 100:126–140.

2. Aron C, et al. 1993. Inherited and environmental determinants of bisexuality in the male rat. In: Haug M, Whalen RE, Aron C, Olsen KL, eds. *The Development of Sex Differences and Similarities in Behavior*. Boston, MA: Kluwer Academic Publishers; 17.

3. Bao AM, Swaab DF. 2010. Sex differences in the brain, behavior, and neuropsychiatric disorders. *The Neuroscientist* 16:550–565.

4. Basson R, Brotto LA. 2003. Sexual psychophysiology and effects of sildenafil citrate in oestrogenised women with acquired genital arousal disorder and impaired orgasm: a randomized controlled trial. *BJOG* 110:1014–1024.

5. Brown WM, et al. 2008. Fluctuating asymmetry and preferences for sex-typical bodily characteristics. *Proc. Natl. Acad. Sci. USA* 105:12938–12943.

6. Brunetti M, et al. 2008. Hypothalamus, sexual arousal and psychosexual identity in human males: a functional magnetic resonance imaging study. *Euro. J. Neurosci.* 27:2922–2927.

7. Castelli MP, et al. 2007. Cannabinoid CB1 receptors in the paraventricular nucleus and central control of penile erection: immunocytochemistry, autoradiography and behavioral studies. *Neuroscience* 147:197–206.

8. Chen H, Ge R-S, Zirkin BR. 2009. Leydig cells: from stem cells to aging. *Mol. Cell. Endocrinol.* 306:9–16.

9. Christensen LW, et al. 1977. Effects of hypothalamic and preoptic lesions on reproductive behavior in male rats. *Brain Res. Bull.* 2:137–141.

10. de Vries GJ, Södersten P. 2009. Sex differences in the brain: the relation between structure and function. *Horm. Behav.* 55:589–596.

11. Edwards DA, et al. 1990. Olfactory bulb removal: effects on sexual behavior and partner-preference in male rats. *Physiol. Behav.* 48:447–450.

12. Forger NG. 2009. The organizational hypothesis and final common pathways: sexual differentiation of the spinal cord and peripheral nervous system. *Horm. Behav.* 55:605–610.

13. Fournier JC, et al. 2010. Antidepressant drug effects and depression severity: a patient-level meta-analysis. *J. Am. Med. Assoc.* 303:47–53.

14. Gahr M, et al. 1998. Sex difference in the size of the neural song control regions in a dueting songbird with similar song repertoire size of males and females. *J. Neuroscience* 18:1124–1131.

15. Gonzalez-Mariscal G, et al. 1994. Participation of opiatergic, GABAergic, and serotonergic systems in the expression of copulatory analgesia in male rats. *Pharmacol. Biochem. Behav.* 49:303–307.

16. Gorski RA, et al. 1978. Evidence for a morphological sex difference within the medial preoptic area of the rat brain. *Brain Res.* 148:333–346.

17. Grus WE, et al. 2005. Dramatic variation of the vomeronasal pheromone receptor gene repertoire among five orders of placental and marsupial mammals. *Proc. Nat. Acad. Sci. USA* 102:5767–5772.

18. Harman SM, et al. 2001. Longitudinal effects of aging on serum total and free testosterone levels in healthy men. Baltimore Longitudinal Study of Aging. *J. Clin. Endocrinol. Metab.* 86:724–731.

19. Hurlemann R, et al. 2010. Oxytocin enhances amygdala-dependent, socially reinforced learning and emotional empathy in humans. *J. Neuroscience* 30:4999–5007.

20. Insel TR. 2010. The challenge of translation in social neuroscience: A review of oxytocin, vasopressin, and affiliative behavior. *Neuron* 65:768–779.

21. Janus SS, Janus CL. 1993. *The Janus Report on Sexual Behavior*. New York: John Wiley & Sons.

22. Kim KW, et al. 2010. CNS-specific ablation of steroidogenic factor 1 results in impaired female reproductive function. *Molec. Endo.* 24:1240–1250.

23. Krapf JM, Simon JA. 2009. The role of testosterone in the management of hypoactive sexual desire disorder in postmenopausal women. *Maturitas* 63:213–219.

24. Lee JW, Erskine MS. 2000. Changes in pain threshold and lumbar spinal cord immediate-early gene expression induced by paced and nonpaced mating in female rats. *Brain Research* 861:26–36.

25. Lourenco D, et al. 2009. Mutations in *NR5A1* associated with ovarian insufficiency. *N. Engl. J. Med.* 360:1200–1210.

26. Mayer-Bahlburg HF, et al. 2008. Sexual orientation in women with classical or non-classical congenital adrenal hyperplasia as a function of degree of prenatal androgen excess. *Arch. Sex Behav.* 37:85–99.

27. Normandin JJ, Murphy A. 2011. Serotonergic lesions of the periaqueductal gray, a primary source of serotonin to the nucleus paragigantocellularis, facilitate sexual behavior in male rats. *Pharmacol. Biochem. Behav.* 98:369–375.

28. Pantages E, Dulac C. 2000. A novel family of candidate pheromone receptors in mammals. *Neuron* 28:835–845.

29. Robinson JE, et al. 2010. Prenatal exposure of the ovine fetus to androgens reduces the proportion of neurons in the ventromedial and arcuate nucleus that are activated by short-term exposure to estrogen. *Biol. Reproduction* 82:163–170.

30. Sahara Y, et al. 2001. Cellular localization of metabotropic glutamate receptors mGluR1, 2/3, 5 and 7 in the main and accessory olfactory bulb of the rat. *Neurosci. Lett.* 312:59–62.

31. Sakamoto H, et al. 2009. Androgen regulates the sexually dimorphic gastrin-releasing peptide system in the lumbar spinal cord that mediates male sexual function. *Endocrinology* 150:3672–3679.

32. Savard G, et al. 2003. Psychiatric aspects of patients with hypothalamic hamartoma and epilepsy. *Epilepsy Disorders* 5:229–234.

33. Savic I, Berglund H. 2010. Androstenol—a steroid derived odor activates the hypothalamus in women. *PLoS ONE* 5:e8651.

34. Sengelaub DR, Forger NG. 2008. The spinal nucleus of the bulbocavernosus: firsts in androgen-dependent neural sex differences. *Horm. Behav.* 53:596–612.

35. Siddle HV, et al. 2010. MHC gene copy number variation in Tasmanian devils: implications for the spread of a contagious cancer. *Proc. Royal Society B* 277:2001–2006.

36. Simmons DA, Hoffman NW, Yahr P. 2011. A forebrain-retrorubral pathway involved in male sex behavior is GABAergic and activated with mating in gerbils. *Neuroscience* 175:162–168.

37. Stender JD, et al. 2010. Genome-wide analysis of estrogen receptor DNA binding and tethering mechanisms identifies Runx1 as a novel tethering factor in receptor-mediated transcriptional activation. *Molec. Cellular Biol.* 30:3943–3955.

38. Stief CG, et al. 1998. The effect of the specific phosphodiesterase (PDE) inhibitors on human and rabbit cavernous tissue in vitro and in vivo. *J. Urology* 159:1390–1393.

39. Wade TD, et al. 2010. Body mass index and breast size in women: same or different genes? *Twin Res. Human Genet.* 13:450–454.

40. Wersinger SR, et al. 2008. Inactivation of the oxytocin and the vasopressin (Avp)1b receptor genes, but not the Avp1a gene, differentially impairs the Bruce effect in laboratory mice. *Endocrinology* 149:116–121.

41. Westberg L, et al. 2009. Influence of androgen receptor repeat polymorphisms on personality traits in men. *J. Psychiatry Neurosci.* 34:205–209.

42. Whalen RE. 1993. Animal sexual differentiation: the early days and current questions. In: Haug M, Whalen RE, Aron C, Olsen KL, eds. *The Development of Sex Differences and Similarities in Behavior.* Boston, MA: Kluwer Academic Publishers; 77–85.

43. Young JK. 1982. A comparison of hypothalami of rats and mice: lack of gross sexual dimorphism in the mouse. *Brain Research* 239:233–239.

44. Young JK. 2010. *Introduction to Cell Biology.* Singapore: World Scientific Press; chap 1.

45. Young JK, et al. 1985. Sex behavior and the sexually dimorphic hypothalamic nucleus in male Zucker rats. *Physiol. Behav.* 36:881–886.

CHAPTER 4

1. Arnulf I, et al. 2008. Kleine-Levin syndrome: a systematic study of 108 patients. *Ann. Neurol.* 63:482–492.

2. Billiard M, Guilleminault C, Dement WC. 1975. A menstruation-linked periodic hypersomnia. Kleine-Levin syndrome or new clinical entity? *Neurology* 25:435–443.

3. Borbely A. 1986. *Secrets of Sleep.* New York: Basic Books.

4. Everson CA. 1995. Functional consequences of sustained sleep deprivation in the rat. *Behav. Brain Res.* 69:43–54.

5. Freud S. (1953 edition). *The Interpretation of Dreams.* Strachey J, trans. New York: Basic Books.

6. Jinka TR, et al. 2011. Season primes the brain in an arctic hibernator to facilitate entrance into torpor mediated by adenosine A_1 receptors. *J. Neuroscience* 31:10752–10758.

7. Krenzer M, et al. 2011. Brainstem and spinal cord circuitry regulating REM sleep and muscle atonia. *PLoS One* 6:e24998.

8. Li KY, et al. 2009. Propofol facilitates glutamatergic transmission to neurons of the ventrolateral preoptic nucleus. *Anesthesiology* 111:1271–1278.

9. Lydic R, et al. 1980. Suprachiasmatic region of the human hypothalamus—homolog to the primate circadian pacemaker. *Sleep* 2:355–361.

10. Nir Y, Tononi G. 2010. Dreaming and the brain: from phenomenology to neurophysiology. *Trends Cogn. Sci.* 14:88–100.

11. Nishino S. 2007. Clinical and neurobiological aspects of narcolepsy. *Sleep Med.* 8:373–399.

12. Oishi Y, et al. 2008. Adenosine in the tuberomammillary nucleus inhibits the histaminergic system via A1 receptors and promotes non-rapid eye movement sleep. *Proc. Natl. Acad. Sci. USA* 105:19992–19997.

13. Okamura H, et al. 2010. Mammalian circadian clock system: molecular mechanisms for pharmaceutical and medical sciences. *Adv. Drug Deliv. Rev.* 62:876–884.

14. Palagini L, Rosenlicht N. 2011. Sleep, dreaming, and mental health: a review of historical and neurobiological perspectives. *Sleep Med. Rev.* 15:179–186.

15. Panda S, et al. 2002. Coordinated transcription of key pathways in the mouse by the circadian clock. *Cell* 109:307–320.

16. Papy JJ, et al. 1982. The periodic hypersomnia and megaphagia syndrome in a young female, correlated with menstrual cycle. *Rev. EEG Neurophys. Clin.* 12:54–61.

17. Passani MB, Blandina P. 2011. Histamine receptors in the CNS as targets for therapeutic intervention. *Trends Pharm. Sci.* 32:242–249.

18. Preston BT, et al. 2009. Parasite resistance and the adaptive significance of sleep. *BMC Evol. Biol.* 9:7–16.

19. Rauchs G, et al. 2011. Sleep contributes to the strengthening of some memories over others, depending on hippocampal activity at learning. *J. Neurosci.* 31:2563–2568.

20. Rolls A, et al. 2010. Sleep and metabolism: role of hypothalamic neuronal circuitry. *Best Pract. Res. Clin. Endocrin. Metab.* 24:817–828.

21. Sachs C, et al. 1982. Menstruation-related periodic hypersomnia: a case study with successful treatment. *Neurology* 32:1376–1379.

22. Sacks O. 1990. *Awakenings*. New York: Vintage.

23. Saper CB, et al. 2001. The sleep switch: hypothalamic control of sleep and wakefulness. *Trends Neurosci.* 24:726–731.

24. Schredl M, et al. 2004. Typical dreams: stability and gender differences. *J. Psychol.* 138:485–494.

25. Shuto H, et al. 2005. Forced exercise-induced flushing of tail skin in ovariectomized mice, as a new experimental model of menopausal hot flushes. *J. Pharmacol. Sci.* 98:323–326.

26. Taheri S, Mignot E. 2002. The genetics of sleep disorders. *Lancet Neurology* 1:242–250.

27. Toh KL, et al. 2001. An h*Per2* phosphorylation site mutation in familial advanced sleep phase syndrome. *Science* 291:1040–1043.

28. Triarhou LC. 2006. The percipient observations of Constantin von Economo on encephalitis lethargica and sleep disruption and their lasting impact on contemporary sleep research. *Brain Res. Bullet.* 69:244–258.

29. Vetrivelan R, et al. 2009. Medullary circuitry regulating rapid eye movement (REM) sleep and motor atonia. *J. Neurosci.* 29:9361–9369.

30. Xi Z, Luning W. 2009. REM sleep behavior disorder in a patient with pontine stroke. *Sleep Med.* 10:143–146.

31. Yamaguchi S, et al. 2003. Synchronization of cellular clocks in the suprachiasmatic nucleus. *Science* 302:1408–1412.

32. Young JK. 1975. A possible neuroendocrine basis of two clinical syndromes: anorexia nervosa and the Kleine-Levin syndrome. *Physiol. Psychol.* 3:322–330.

33. Young JK, Polston E. 2012. Specialized features of the arcuate nucleus of the hypothalamus. In: Dudas B, ed. *The Human Hypothalamus: Anatomy, Functions, and Disorders.* New York: Nova Science Publishers.

34. Young T, et al. 1993. The occurrence of sleep-disordered breathing among middle-aged adults. *New Eng. J. Med.* 328:1230–1235.

35. Young JK, et al. 2005. Orexin stimulates breathing via medullary and spinal pathways. *J. Appl. Physiol.* 98:1387–1395.

36. Zhang S, et al. 2007. Sleep/wake fragmentation disrupts metabolism in a mouse model of narcolepsy. *J. Physiol.* 581:649–663.

CHAPTER 5

1. Anderson E, Haymaker W. 1974. Breakthroughs in hypothalamic and pituitary research. *Prog. Brain Res.* 41:1–60.

2. Ara T, et al. 2003. Impaired colonization of the gonads by primordial germ cells in mice lacking a chemokine, stromal cell-derived factor-1 (SDF-1). *Proc. Natl. Acad. Sci. USA* 100:5319–5323.

3. Backholer K, et al. 2010. Kisspeptin cells in the ewe brain respond to leptin and communicate with neuropeptide Y and proopiomelanocortin cells. *Endocrinology* 151:2233–2243.

4. Bailar JC, Gornik HL. 1997. Cancer undefeated. *N. Eng. J. Med.* 336:1569–1574.

5. Bennett PH, et al. 2003. Epidemiology of diabetes mellitus. In: Porte D, Sherwin RS, Baron A, eds. *Ellenberg and Rifkin's Diabetes Mellitus.* 6th ed. New York: McGraw-Hill.

6. Briski KP. 1998. Glucoprivic induction of fos immunoreactivity in hypothalamic dopaminergic neurons. *Neuroreport* 9:289–295.

7. Burcelin R, Thorens B. 2001. Evidence that extrapancreatic GLUT2-dependent glucose sensors control glucagon secretion. *Diabetes* 50:1282–1289.

8. Coppari R, et al. 2005. The hypothalamic arcuate nucleus: a key site for mediating leptin's effects on glucose homeostasis and locomotor activity. *Cell Metab.* 1:63–72.

9. Coughlin SS, et al. 2004. Diabetes mellitus as a predictor of cancer mortality in a large cohort of US adults. *Am. J. Epidemiol.* 159:1160–1167.

10. Daughaday WH. 2006. Endocrinology—the way we were: a personal history of somatomedin. *Growth Horm. IGF Res.* 16(suppl A):S305.

11. Dilman V. 1994. *Development, Aging, and Disease: A New Rationale for an Intervention Strategy.* Young JK, trans. New York: Gordon & Breach; 384.

12. Downs JL, Wise PM. 2009. The role of the brain in female reproductive aging. *Mol. Cell. Endo.* 299:32–38.

13. Fielding JE. 1999. Public health in the twentieth century: advances and challenges. *Annu. Rev. Public Health* 20:xiii–xxx.

14. Fujikawa T, et al. 2010. Leptin therapy improves insulin-deficient type 1 diabetes by CNS-dependent mechanisms in mice. *Proc. Natl. Acad. Sci. USA* 107:17391–17396.

15. Gallagher EJ, Leroith D. 2011. Minireview: IGF, insulin, and cancer. *Endocrinology* 152:2546–2551.

16. Grosvenor AE, Laws ER. 2008. The evolution of extracranial approaches to the pituitary and anterior skull base. *Pituitary* 11:337–345.

17. Guillemin R. 1975. Pioneering in neuroendocrinology. In: Meites J, et al., eds. *Pioneers in Neuroendocrinology.* New York: Plenum Press.

18. Haber DA, et al. 2011. The evolving war on cancer. *Cell* 145:19–25.

19. Harvey PW. 2012. Hypothesis: prolactin is tumorigenic to human breast: dispelling the myth that prolactin-induced mammary tumors are rodent-specific. *J. Appl. Toxicol.* 32:1–9.

20. Hu L, et al. 2008. Converse regulatory functions of estrogen receptor-a and -b subtypes expressed in hypothalamic gonadotropin-releasing hormone neurons. *Molec. Endocrinol.* 22:2250–2259.

21. Kaplan SA. 2007. The pituitary gland: a brief history. *Pituitary* 10:323–325.

22. Kriegsfeld LJ, et al. 2006. Identification and characterization of a gonadotropin-inhibitory system in the brains of mammals. *Proc. Natl. Acad. Sci. USA* 103:2410–2414.

23. Lakka TA, et al. 2004. Leptin and leptin receptor gene polymorphisms and changes in glucose homeostasis in response to regular exercise in nondiabetic individuals: the HERITAGE family study *Diabetes* 53:1603–1608.

24. Lamberts SWJ, et al. 1997. The endocrinology of aging. *Science* 278:419–424.

25. Lee JH, et al. 1996. KiSS-1, a novel human malignant melanoma metastasis-suppressor gene. *J. Natl. Cancer Inst.* 88:1731–1737.

26. Marty N, et al. 2005. Regulation of glucagon secretion by glucose transporter type 2 (glut2) and astrocyte-dependent glucose sensors. *J. Clin. Invest.* 115:3545–3550.

27. Medvei CM. 1982. *A History of Endocrinology.* Boston, MA: MTP Press.

28. Millan C, et al. 2010. Glial glucokinase expression in adult and post-natal development of the hypothalamic region. *ASN Neuro.* 3:e00035.

29. Niikura Y, et al. 2009. Aged mouse ovaries possess rare premeiotic germ cells that can generate oocytes following transplantation into a young host. *Aging* 1:971–976.

30. Olefsky JM, Kruszynska YT. 2003. Insulin resistance. In: Porte D, Sherwin RS, Baron A, eds. *Ellenberg and Rifkin's Diabetes Mellitus.* 6th ed. New York: McGraw-Hill.

31. Payami H, et al. 1996. Gender difference in apolipoprotein E-associated risk for familial Alzheimer disease: a possible clue to the higher incidence of Alzheimer disease in women. *Am. J. Human Genet.* 58:803–811.

32. Quennell JH, et al. 2009. Leptin indirectly regulates gonadotropin-releasing hormone neuronal function. *Endocrinology* 150:2805–2812.

33. Rajkovic A, et al. 2004. NOBOX deficiency disrupts early folliculogenesis and oocyte-specific gene expression. *Science* 305:1157–1159.

34. Rance NE. 2009. Menopause and the human hypothalamus: evidence for the role of kisspeptin-neurokinin B neurons in the regulation of estrogen negative feedback. *Peptides* 30:111–122.

35. Razavi R, et al. 2006. TRPV1+ sensory neurons control cell stress and islet inflammation in autoimmune diabetes. *Cell* 127:1123–1135.

36. Schipper HM, et al. 1998. The peroxidase-positive subcortical glial system. In: Schipper HM, ed. *Astrocytes in Brain Aging and Neurodegeneration*. Georgetown TX: RG Landes Co.

37. Schmidt SP, et al. 2010. Misfolding of short-chain acyl-CoA dehydrogenase leads to mitochondrial fission and oxidative stress. *Mol. Genet. Metab.* 100:155–162.

38. Shuto H, et al. 2005. Forced exercise-induced flushing of tail skin in ovariectomized mice, as a new experimental model of menopausal hot flushes. *J. Pharmacol. Sci.* 98:323–326.

39. Telser A, Young JK, Baldwin K. 2007. *Elsevier's Integrated Histology*. Philadelphia, PA: Elsevier; 394.

40. Tkacs NC, et al. 2000. Presumed apoptosis and reduced arcuate nucleus neuropeptide Y and pro-opiomelanocortin mRNA in non-coma hypoglycemia. *Diabetes* 49:820–826.

41. Tsutsui K, et al. 2010. Discovery and evolutionary history of gonadotrophin-inhibitory hormone and kisspeptin: new key neuropeptides controlling reproduction. *J. Neuroendocrinol.* 22:716–727.

42. van Oostrom AJHHM, et al. 2002. Insulin resistance and vessel endothelial function. *J. Royal Soc. Med.* 95(Suppl 12): 54–58.

43. World Health Organization Fact Sheet #310, 2008.

44. Yamanaka K, et al. 2008. Astrocytes as determinants of disease progression in inherited amyotrophic lateral sclerosis. *Nat. Neurosci.* 11:251–253.

45. Young JK. 2002. Anatomical relationship between specialized astrocytes and leptin-sensitive neurons. *J. Anatomy* 201:85–90.

46. Young JK, Baker JH, Montes MI. 2000. The brain response to 2-deoxy glucose is blocked by a glial drug. *Pharmacol. Biochem. Behav.* 67:233–239.

47. Young JK, Baker JH, Muller T. 1996. Immunoreactivity for brain fatty acid binding protein in Gomori-positive astrocytes. *Glia* 16:218–226.

48. Young JK, McKenzie JL. 2004. GLUT2 immunoreactivity in Gomori-positive astrocytes of the hypothalamus. *J. Histochem. Cytochem.* 52:1519–1524.

49. Young JK, et al. 1990. Association of iron-containing astrocytes with dopaminergic neurons of the arcuate nucleus. *J. Neurosci. Res.* 25:204–213.

50. Zierath JR, et al. 1998. Evidence against a direct effect of leptin on glucose transport in skeletal muscle and adipocytes. *Diabetes* 47:1–4.

CHAPTER 6

1. Bittman M. 2009. *Food Matters: A Guide to Conscious Eating*. New York: Simon & Schuster.

2. Lamberg L. 1999. Researchers urged to tell public how animal studies benefit human health. *J. Am. Med. Assoc.* 282:619–621.

3. Pauly D, et al. 2002. Towards sustainability in world fisheries. *Nature* 418:689–695.

4. Pepperberg I. 2008. *Alex & Me: How a Scientist and a Parrot Discovered a Hidden World of Animal Intelligence—and Formed a Deep Bond in the Process.* New York: HarperCollins.

Index

About the Author

John Young is currently professor in the Department of Anatomy at Howard University College of Medicine in Washington, D.C. He received a Kaiser-Permanente Award for Excellence in Teaching in 1998. He has performed research on a portion of the brain called the hypothalamus for 35 years. Dr. Young has published a number of textbooks (*Elsevier's Integrated Histology*, *Wheater's Review of Histology and Basic Pathology*, and *Introduction to Cell Biology*) and has authored 45 scientific papers.